Springer
Berlin
Heidelberg
New York
Barcelona
Budapest
Hong Kong
London
Milan
Paris
Santa Clara
Singapore
Tokyo

Richard M. Eglen (Ed.)

5-HT$_4$ Receptors in the Brain and Periphery

 Springer

Richard M. Eglen, Ph.D.

Roche Bioscience
Palo Alto, California, U.S.A.

QP
801
.S4
A15
1998

ISBN: 3-540-64164-5 Springer-Verlag Berlin Heidelberg New York
Biotechnology Intelligence Unit

Library of Congress Cataloging-in-Publication data

5-HT4 receptors in the brain and periphery / [edited by] Richard M. Eglen.
  p. cm. — (Biotechnology intelligence unit)
  Includes bibliographical references and index.
  ISBN 1-57059-521-6 (alk. paper); 3-540-64164-5 (alk.paper)
    1. Serotonin—Receptors. 2. Neurotransmitter receptors. I. Eglen, Richard M. II. Series.
    [DNLM: 1. Serotonin—metabolism. 2. Receptors, Serotonin—metabolism. 3. Serotonin
Antagonists—pharmacology. 4. Serotonin Agonists—pharmacology. 5. Nervous system—
drug effects. QV 126 Z9942 1998]
QP801.S4A15 1998
612.8—dc21
DNLM/DLC   97-46652
for Library of Congress   CIP

Typesetting: R.G. Landes Company, Georgetown, TX, U.S.A.

SPIN 10670718   31/3111 - 5 4 3 2 1 0 - Printed on acid-free paper

PREFACE

Receptors for 5-hydroxytryptamine (5-HT) exist in at least 14 different subtypes, each of which exhibit a unique pharmacology, distribution and function. In recent years, the 5-HT_4 receptor, has been one of the most intensively studied, both from a basic research standpoint and as a target for novel therapeutics. Indeed, the receptor is now very well characterized on the basis of structural and operational criteria, reflecting the high level of interest. As a result, several physiological roles of the receptor are now emerging.

The 5-HT_4 receptor is a seven transmembrane spanning G-protein coupled receptor, that in all tissues studied to date, positively couples to adenylate cyclase. The receptor exists in at least two isoforms (5-HT_{4S} and 5-HT_{4L}) that differ in their distribution in both central and peripheral nervous systems. Pharmacologically, the 5-HT_4 receptor is defined by several selective agonists and antagonists, some with good bioavailability in a number of species. Selective radioligands are now available for the receptor and these have facilitated the mapping of 5-HT_4 distribution in the brain.

Based upon the anatomical location and activity of selective ligands in animal models, emerging data suggest that 5-HT_4 receptors play a role in learning and memory formation. In the periphery, several lines of evidence indicate that the receptor modulates gastrointestinal motility and secretion, increases heart rate, enhances steroid secretion from the adrenal gland and augments bladder emptying.

A number of 5-HT_4 receptor agonists, with varying degrees of selectivity, are in clinical development, primarily in terms of their potential to augment gastric motility. Although clinical data is unavailable, it is possible that selective 5-HT_4 receptor antagonists will have therapeutic utility in the treatment of irritable bowel syndrome, urge incontinence or cardiac arrythmias.

The intent of this book is to bring together, in a single volume, a series of chapters covering all the above aspects of the receptor. Each of these have been written by an acknowledged expert in the field and collectively present a current synopsis of 5-HT_4 receptor research. Given the evident progress made in this area in the last decade, one may look forward to an equally exciting era of progress in years to come. Moreover, as clinical experience with selective 5-HT_4 receptor ligands increases, important therapeutics will undoubtedly emerge.

CONTENTS

EDITOR

Richard M. Eglen, Ph.D.
Roche Bioscience
Palo Alto, California, U.S.A.

CONTRIBUTORS

Nika Adham, Ph.D.
Synaptic Pharmaceutical
  Corporation
Paramus, New Jersey, U.S.A.
Chapter 2

Rodrigo Andrade, Ph.D.
Dept. of Psychiatry and
  Behavioral Neurosciences
and
Cellular and Clinical
  Neurobiology Training
  Program
Wayne State University School
  of Medicine
Detroit, Michigan, U.S.A.
Chapter 4

Janine M. Barnes, Ph.D.
Department of Pharmacology
The Medical School
University of Birmingham
Edgbaston, Birmingham, U.K.
Chapter 5

Nicholas M. Barnes, Ph.D.
Department of Pharmacology
The Medical School
University of Birmingham
Edgbaston, Birmingham, U.K.
Chapter 5

Joël Bockaert, Ph.D.
Centre National de la Recherche
  Scientifique
Cedex, France
Chapter 3

Theresa A. Branchek, Ph.D.
Synaptic Pharmaceutical
  Corporation
Paramus, New Jersey, U.S.A.
Chapter 2

Esther Chapin, B.A.
Dept. of Psychiatry and
  Behavioral Neurosciences
and
Cellular and Clinical
  Neurobiology Training
  Program
Wayne State University School
  of Medicine
Detroit, Michigan, U.S.A.
Chapter 4

Robin D. Clark, Ph.D.
Roche Bioscience
Palo Alto, California, U.S.A.
Chapter 1

Vincent Contesse, Ph.D.
Laboratory of Cellular and
  Molecular Neuroendocrinology
University of Rouen
Rouen, France
Chapter 9

Catherine Delarue, Ph.D.
Laboratory of Cellular and
  Molecular Neuroendocrinology
University of Rouen
Rouen, France
Chapter 9

Aline Dumuis, Ph.D.
Centre National de la Recherche
    Scientifique
Cedex, France
Chapter 3

Anthony P.D.W. Ford, Ph.D.
Urogenital Pharmacology
Center for Biological Research
Roche Bioscience
Palo Alto, California, U.S.A.
Chapter 8

Christophe Gerald, Ph.D.
Synaptic Pharmaceutical
    Corporation
Paramus, New Jersey, U.S.A.
Chapter 2

Samir Haj-Dahmane, Ph.D.
Department of Psychiatry and
    Behavioral Neurosciences
and
Cellular and Clinical Neurobiology
    Training Program
Wayne State University School
    of Medicine
Detroit, Michigan, U.S.A.
Chapter 4

Sharath S. Hegde, Ph.D.
Roche Bioscience
Palo Alto, California, U.S.A.
Chapter 7

Alberto J. Kaumann, Ph.D.
The Babraham Institute
Human Pharmacology Laboratory
Babraham, Cambridge, U.K.
Chapter 6

M. Shannon Kava, B.A., M.S.
Urogenital Pharmacology
Center for Biological Research
Roche Bioscience
Palo Alto, California, U.S.A.
Chapter 8

Jean-Marc Kuhn, Ph.D.
Department of Endocrinology
Group for Hormone Research
European Institute for Peptide
    Research
University Hospital of Rouen
Rouen, France
Chapter 9

Hervé Lefebvre, M.D., Ph.D.
Department of Endocrinology
Group for Hormone Research
European Institute for Peptide
    Research
University Hospital of Rouen
Rouen, France
Chapter 9

Louise Sanders, Ph.D.
Department of Plant Sciences
Downing Site
Cambridge, U.K.
Chapter 6

Gareth J. Sanger, Ph.D.
Neurology Research Department
SmithKline Beecham
    Pharmaceuticals
Harlow, Essex, U.K.
Chapter 10

Hubert Vaudry, M.D., Ph.D.
Laboratory of Cellular and
    Molecular Neuroendocrinology
University of Rouen
Rouen, France
Chapter 9

GLOSSARY

Chapter 4

AHP   The afterhyperpolarization produced by the opening of calcium activated potassium channels following calcium influx into the cell during an action potential.

I_h A voltage-sensitive cation nonselective current that activates upon hyperpolarization in the subthreshold range.

PKA   The family of protein kinases activated by cAMP.

PKI   A peptide inhibitor of PKA.

CHAPTER 1

Medicinal Chemistry of 5-HT$_4$ Receptor Ligands

Robin D. Clark

Introduction

The discovery and classification of the 5-HT$_4$ receptor was greatly facilitated by identification of ligands that interacted with this receptor and that provided tools for the elucidation of receptor signaling and function. Thus tropisetron (ICS-205 930, 1*) was found to be a weak antagonist of 5-HT induced stimulation of adenylate cyclase activity in primary cultures of mouse embryo colliculi neurones with an apparent inhibition constant (K$_i$) of 997 nM.[1] Tropisetron was also reported to inhibit 5-HT stimulated adenylate cyclase activity in guinea pig hippocampus with a pK$_i$ of 7.6,[2] and the connection thereby established between the 5-HT receptor responsible for the observed activity in these two CNS preparations was largely responsible for the proposed classification of the 5-HT$_4$ receptor (for reviews see refs. 3-5). The subsequent finding that certain members of the benzamide family of gastrointestinal prokinetic agents were agonists of the 5-HT$_4$ receptor in mouse embryo colliculi neurones[6] and in guinea pig hippocampus[7] represented the first step in the functional characterization of this receptor. The benzamides used in these seminal studies were, in decreasing order of potency, cisapride (2), renzapride (3), zacopride (4), BRL 20627 (5) and the progenitor of the benzamide class metoclopramide (6). Whereas tropisetron and the benzamides played a pivotal role in establishing the existence of

---

Bold numbers in text indicate structures, which are illustrated at the end of the chapter along with all figures and tables.

5-HT$_4$ Receptors in the Brain and Periphery, edited by Richard M. Eglen.
© 1998 Springer-Verlag and R.G. Landes Company.

the 5-HT$_4$ receptor, the utility of these drugs in the further characterization of the pharmacology of this receptor was severely limited by lack of selectivity. Tropisetron, renzapride and zacopride are high affinity 5-HT$_3$ receptor antagonists,[8] and cisapride and metoclopramide have affinity for multiple receptor types.[9] This chapter describes the significant progress made in the medicinal chemistry of 5-HT$_4$ receptor ligands subsequent to the identification of these early leads which has resulted in the discovery of potent and selective 5-HT$_4$ receptor agonists and antagonists, several of which are currently undergoing clinical evaluation. A previous review has admirably covered certain aspects of the medicinal chemistry of 5-HT$_4$ receptor antagonists, most notably the structural relationships between 5-HT$_3$ and 5-HT$_4$ receptor antagonists.[10]

Pharmacological Evaluation of 5-HT$_4$ Receptor Ligands

In addition to the mouse colliculi neuron cAMP assay,[1] other in vitro methods for determining 5-HT$_4$ receptor agonist and antagonist activity were developed using preparations from guinea pig and rat, and the latter systems have been most extensively employed for the characterization of the ligands described in this chapter. In the guinea pig ileum, the ability of test compounds to enhance the twitch (contractile) response evoked by electrical field stimulation of the longitudinal muscle myenteric plexus has been used to determine 5-HT$_4$ receptor agonist activity.[11] Antagonist activity has been measured as a function of the inhibition of 5-HT-induced contractions. The guinea pig isolated distal colon longitudinal muscle myenteric plexus, in which 5-HT elicits a contractile response, has also been used to characterize agonists and antagonists.[12] Another widely used preparation has been the rat esophageal tunica muscularis mucosae, in which 5-HT causes relaxation of carbachol-induced tension.[13]

With the advent of selective radioligands for the 5-HT$_4$ receptor, binding assays have supplemented the in vitro functional test systems. Tritiated GR 113808, a high affinity 5-HT$_4$ receptor antagonist, has been widely used for radioligand binding assays in guinea pig striatum and hippocampus and in rat striatum.[14] Radiodinated SB 207710 is another high affinity antagonist which has been used to label 5-HT$_4$ receptors in piglet caudate and hippocampus.[15] The structures of these radioligands are presented in later sections.

A variety of in vivo models have been used to evaluate the functional effects of 5-HT$_4$ receptor ligands and to determine the potential therapeutic utilities of these agents. These have included models of gastrointestinal motility in rat[16] and mouse,[17] in which 5-HT$_4$ agonists have a prokinetic effect. The Heidenhain pouch model in the conscious dog, wherein 5-HT$_4$ receptors mediate cholinergically driven contraction, has been used to evaluate 5-HT$_4$ antagonists.[18] In conscious mice, inhibition of 5-hydroxytryptophan-induced diarrhea has also been used to assess the effect of 5-HT$_4$ antagonists on gastrointestinal function.[19,20] In the anesthetized micropig, the effect of 5-HT$_4$ receptor ligands on heart rate has afforded a useful measure of 5-HT$_4$ agonist (tachycardia) and antagonist (inhibit 5-HT-induced tachycardia) properties.[21,22] In the CNS realm, rodent models of spatial learning and memory have been used to evaluate 5-HT$_4$ agonists for procognitive activity.[23]

Structure-Activity Relationships of 5-HT$_4$ Receptor Ligands

Shortly after the classification of the 5-HT$_4$ receptor, a number of ligands with somewhat improved affinity and selectivity over the initial leads 1-6 were reported. An interesting early finding was that the ester corresponding to metoclopramide, SDZ 205-557 (7), was a moderately potent 5-HT$_4$ antagonist (pA$_2$ 7.4),[24] although this compound demonstrated only marginal selectivity (ca. 3-fold) over the 5-HT$_3$ receptor.[22] The related esters LY 297524 (8, pA$_2$ 7.8)[25] and RS-23597 (9, pA$_2$ 7.8)[26] were also reported to be 5-HT$_4$ antagonists with RS-23597 demonstrating improved selectivity (125-fold) over the 5-HT$_3$ receptor. Significantly higher potency and selectivity were achieved with the synthesis of the indole ester GR 113808 (10, pA$_2$ 9.2)[27] and in particular, benzoate SB 204070 (11, pA$_2$ 10.8).[28] All of these antagonists are esters and, as such, have limited utility for in vivo studies because of hydrolytic instability. Thus a major challenge facing medicinal chemists became the development of 5-HT$_4$ receptor antagonists with pharmacokinetic properties suitable for systemic, preferably oral, administration.

Whereas antagonists with improved 5-HT$_4$ receptor affinity and selectivity as compared to the initial lead tropisetron (1) were rapidly discovered, agonists with improved activity and selectivity relative to benzamides 2-6 proved to be somewhat more elusive. The benzimidazolone BIMU 8 (12, pEC$_{50}$ 7.14, efficacy 124% of 5-HT)[29] was shown to have agonist properties essentially equal to those of cisapride (**2**) in the stimulation of adenylate cyclase in mouse embryo colliculi neurons; however, the high affinity of this ligand for the 5-HT_3 receptor (pK_i 8.7) compromised its utility for 5-HT_4 receptor agonist studies. The benzimidazolone series provided an interesting structure-activity study as the unsubstituted parent DAU 6215 (**13**, pEC_{50} 5.58) was found to be essentially devoid of 5-HT_4 receptor agonist activity, whereas the N-ethyl analog BIMU 1 (**14**, pEC_{50} 6.44, efficacy 72% of 5-HT) was a partial agonist.[30] A related derivative, DAU 6285 (**15**) was a 5-HT_4 receptor antagonist with a pA_2 of 6.5-7.1 depending on the experimental model.[31] Although not a potent 5-HT_4 receptor antagonist, DAU 6285 did possess a modicum of selectivity over the 5-HT_3 receptor (pK_i 6.5). Lack of selectivity over the 5-HT_3 receptor, as found for the BIMU series of 5-HT_4 receptor agonists, would be noted in other series and provide an obstacle in the quest for ligands with selective 5-HT_4 agonist properties.

The following sections deal with structure-activity relationships (SAR) of the more selective and/or more potent 5-HT_4 receptor ligands reported through mid 1997. Agonists are dealt with first and the discussion is divided according to chemical class. Subsequently, antagonists are described, again according to chemical class. As is evident from the preceding paragraphs, there can be a close structural similarity within certain classes of 5-HT_4 receptor agonists and antagonists; hence, there is inevitably some degree of overlap between the sections. For ease of comparison, results of functional and binding affinity determinations are, with some exceptions, presented as pEC_{50}, pIC_{50} and pK_i values with conversions made where required.

5-HT₄ Receptor Agonists

Benzoates and Benzamides

Structure-activity relationship studies related to the 5-HT_4 receptor antagonist SDZ 205-557 (**7**) carried out independently by three research groups indicated that the corresponding 2-(1-piperidinyl)ethyl benzoate **16** possessed 5-HT_4 receptor agonist activity,[32-34] and this compound was designated the code name ML 10302.[32] ML 10302 contracted quiescent guinea pig ileum with a pA_2 of 8.4 with a maximum efficacy 44% that of 5-HT[34] and caused relaxation of carbachol-contracted rat esophagus with pEC_{50} and % efficacy values in the ranges of 7.7-8.6 and 50-80%, respectively.[32-35] In ligand bind-

ing studies, **16** had high affinity for the 5-HT$_4$ receptor in rat striatum (pK$_i$ 9.0) and greater than 700-fold selectivity over the rat posterior cortex 5-HT$_3$ receptor.[35] Binding affinity for the guinea pig
striatum 5-HT$_4$ receptor was considerably lower (pK$_i$ 7.9).[32]

Monosubstitution in the piperidine ring of **16** with methyl or
small polar substituents led to derivatives (**17-23**) that maintained
the high affinity and partial agonist properties of the parent (Table
1.1).[36] Replacement of the piperidine ring with other cyclic or bicy-
clic amines also resulted in maintained 5-HT$_4$ receptor partial agonist activity, although the parent **16** remained the highest affinity,
most efficacious agonist in the series (Table 1.2). It was interesting to
note that quaternary nitrogen derivatives (**29, 30**) were also active.

Somewhat surprisingly, the 3,5-dimethylpiperidinyl analogs **31**
and **32** were found to be high affinity 5-HT$_4$ antagonists.[36] The *cis*-
analog **31** had a pK$_i$ of 9.6 in the rat striatum, and functional 5-HT$_4$
receptor antagonist activity was demonstrated in the guinea pig ileum
(pA$_2$ 8.0) and rat esophagus (pA$_2$ 8.6). The *trans*-derivative **32** had
slightly lower activity in the three assays (pK$_i$ 8.6, pA$_2$ 7.9 in guinea
pig ileum and 8.2 in rat esophagus). A hypothesis was presented to
explain the different profiles observed for **16** and **31,32** in which the
agonist (**16**) was proposed to adopt a folded conformation to present
the lone pair of the basic nitrogen atom in a particular orientation
required for binding to, and activation of the receptor. Ligands **31**
and **32** bind to the receptor, but steric hindrance induced by the di-
methyl substituents could prevent the adoption of the folded conformation presumed to be responsible for agonist activity.

The importance of the ester linkage to the 5-HT$_4$ agonist properties of **16** was underscored by the relative lack of activity of the
bioisosteric amide (**33**)[33,34,36] and ketone (**34**)[33,36] derivatives; both
were two log-orders lower in affinity for the rat striatum 5-HT$_4$ receptor with pK$_i$ values of 6.9 and 7.0, respectively, and neither demonstrated agonist activity.

On account of their susceptibility to hydrolysis, esters such as
16 were expected to have limited utility for in vivo applications, hence
5-HT$_4$ receptor agonists with increased stability, primarily
benzamides, were prepared and evaluated. Prior to the classification
of the 5-HT$_4$ receptor, benzamides related to metoclopramide (**6**), as
exemplified by BRL 20627 (**5**) and renzapride (**3**), had been synthesized in a program to develop improved gastrointestinal
prokinetic agents.[37,38] With the discovery that these agents possessed

5-HT$_4$ receptor agonist properties,[6] it became possible to expedite further work on this series by use of functional and in vivo 5-HT$_4$ receptor based assays. Earlier members of the BRL 20627 series that showed gastric prokinetic activity in the rat comparable or slightly better than the parent (5) included the 5,6-diazabicylic derivative 35 and the 2-oxa-substituted analogs 36 and 37.[39] Although 5-HT$_4$ receptor functional or binding data on these compounds were not given, on the basis of in vivo activity they would be expected to have 5-HT$_4$ agonist properties in the same range as BRL 20627 (pEC$_{50}$ 5.49, 60 % efficacy 5-HT in mouse colliculi neurons).[6]

Significantly more potent gastric prokinetic agents such as renzapride (3) were subsequently prepared, and a series of related azabicyclo derivatives were evaluated for 5-HT$_4$ receptor agonist activity in functional and in vivo assays (Table 1.3).[40] Several interesting facets of SAR resulted from this study, including the finding that the equatorial (*endo*) isomers were more potent 5-HT$_4$ receptor agonists than their axial (*exo*) counterparts, a result opposite of that previously observed in the BRL 20627 quinolizidine series. Also noteworthy was that the enantiomers of renzapride (3) had essentially the same activity as the racemate (data not shown), implying a lack of enantioselectivity in the region in which the 3-carbon bridge of 3 binds to the 5-HT$_4$ receptor. In accordance with this finding was the high activity of the azatricyclic compound 42 which incorporates carbon-bridges in both regions expected to be occupied (separately) by the 3-carbon bridges in the renzapride enantiomers. Additionally, it was observed that relatively minor changes in the renzapride structure led to an apparent increase in selectivity for the 5-HT$_4$ receptor vs. the 5-HT$_3$, e.g., the 1-carbon deletion in the azabicyclic bridge to give derivative 40.

Another modification of renzapride that afforded more potent and selective 5-HT$_4$ agonist properties was quaternization of the side-chain basic nitrogen with butyl bromide.[41] Quaternary derivative 38 (SB 205149) had a pEC$_{50}$ value of 8.0 and an intrinsic activity 0.8 that of 5-HT in the rat esophagus and a pK$_i$ of 6.9 for binding the rat cortex 5-HT$_3$ receptor. Thus 38 was 6-fold more potent than renzapride as a 5-HT$_4$ receptor agonist with a ca. 25-fold decrease in affinity for the 5-HT$_3$ receptor.

Another study of benzamides in which the side-chain basic nitrogen was constrained within rigid aza-cyclic systems resulted in the identification of the potent 5-HT$_4$ receptor agonist SC-53116

(43).[9,42] Taking into account the low agonist potency of 2-methyl-5-HT, hypothetical bioactive conformations for 5-HT were generated by molecular modeling. In these conformations, the distance between the presumed key points of interaction with the receptor, i.e., the basic nitrogen and the lone pair of the phenolic hydroxyl group, was determined to be 7.6-8.0 Å (Fig. 1.1).[42] In the benzamides the "virtual ring" provided by the intramolecular hydrogen bond between the *ortho*-methoxy group and the amide NH imparts structural rigidity similar to that of the indole pyrrole ring. It was therefore suggested that these key interactions were mimicked by a lone pair of the benzamide carbonyl group and a suitably disposed side chain basic nitrogen, as in the aza-adamantane **39** (Fig. 1.2). It was noted that in weak 5-HT₄ receptor agonists such as metoclopramide, the lone pair-nitrogen distance is shorter (ca. 7.1 Å) than that predicted by the model.

On the basis of this analysis, a rigorous systematic study of aza-tricyclic and aza-bicyclic ring systems was carried out, including enantiomeric pairs where appropriate.[42] Salient features of the results of this study are presented in Table 1.4. Aza-adamantane scaffolds as in compounds **39** and **40** provided initial encouraging results[43] which prompted the subsequent synthesis and evaluation of aza-noradamantanes (e.g., **41**) and bridged pyrrolizidines (e.g., **42**). Excision of the methano bridge from **42** provided *endo*-substituted pyrrolizidine **43** (SC-53116), the most potent agonist in the series. It is of interest to note that the lone pair-N distance in **43** is essentially the same (8.0 Å) as that predicted for 5-HT (Fig. 1.3). Structure-activity studies in the pyrrolizidine series indicated enantiospecificity in the agonist activity of **43** (vs. enantiomer **44**) and the importance of the *endo*-substitution pattern (vs. *exo*-isomer **45**). The importance of the 1-carbon tether between the pyrrolizidine ring and the amide nitrogen was emphasized by decreased activity of directly linked analogs such as **46**. The ester analog of SC-53316 (SC-55822) was a high affinity partial agonist with a pEC_{50} of 1.5 nM (57% efficacy of 5-HT) and a binding pK_i of 9.5.

SC-53116 was nearly a full agonist in the rat esophagus (93% of the maximum efficacy of 5-HT)[9] and demonstrated moderate selectivity (ca. 10- to 30-fold depending on the assay) over the 5-HT₃ receptor.[9,42] The compound had virtually no affinity for a range of other serotonergic, dopaminergic and adrenergic receptors. The racemic mixture **43,44** (known as SC-49518) was found to enhance gastric emptying and to stimulate gastrointestinal motility in vivo in the dog by a mechanism involving 5-HT$_4$ receptor stimulation.[44]

The effect of locking the conformation of the basic nitrogen containing side chain on 5-HT$_4$ (and 5-HT$_3$) receptor affinity was also investigated by synthesis of the macrocyclic benzamide **47**.[45] This interesting molecule was designed to provide a rigid framework that would essentially fix the orientation of the nitrogen lone pair, a key determinant in binding to 5-HT receptors. In ligand binding studies, compound **47** had a pK$_i$ of 7.5 at the 5-HT$_4$ receptor in rat striatum with slightly lower affinity for the 5-HT$_3$ receptor (pK$_i$ 7.3). In the electrically-stimulated guinea pig ileum, **47** was found to be a partial agonist with a pEC$_{50}$ of 7.2 (efficacy 69% 5-HT). The 5-HT$_4$ receptor affinity of **47** was significantly higher than that of the less constrained piperidine congener **48** (pK$_i$ 6.1), but only marginally higher than the potent 5-HT$_3$ ligand BRL 24682 (**49**, 5-HT$_4$ pK$_i$ 7.3, 5-HT$_3$ pK$_i$ 9.1). With the aid of molecular modeling, it was concluded that **49** could adopt a conformation in which the nitrogen lone pair had the same orientation as in macrocycle **47**, hence the similar affinity of the two ligands in binding to the 5-HT$_4$ receptor. On the other hand, the higher affinity of **49** for the 5-HT$_3$ receptor was rationalized by invoking a different conformation with a resultant lone pair orientation not available to the rigid structure **47**.

The effect of converting the "virtual ring" formed by the intramolecular hydrogen bond in the 5-HT$_3$/5HT$_4$ ligand zacopride (**4**) into an actual ring was studied through the agency of the enantiomeric quinuclidinyl naphthalimides (S)- and (R)-**50** (RS-56532) (Table 1.5).[46] It was found that (R)-**50** had higher affinity for the 5-HT$_3$ receptor than antipode (S)-**50**, whereas the converse was true in binding to the 5-HT$_4$ receptor. Furthermore, in vitro evaluation revealed that only (S)-**50** had measurable 5-HT$_4$ receptor agonist properties, being a potent agonist in the rat esophagus. Taken together, these results presented an intriguing dichotomy when compared to the corresponding receptor interactions of the enantiomers of zacopride (**4**), wherein (S)-zacopride had higher affinity for both the 5-HT$_3$ and 5-HT$_4$ receptor. Thus there was a reversal in the enantioselectivity of binding to the 5-HT$_3$ receptor, but not to the 5-HT$_4$ receptor, for the two pairs of enantiomers.

The enantiomers of RS-56532 afforded a useful pair of tools for the evaluation of 5-HT$_4$ mediated responses as compared to those mediated by the 5-HT$_3$ receptor and a full account of their in vitro pharmacology was published.[47] Further SAR work in this series provided the achiral analog **51** (RS-66331) which retained a profile similar to that of (S)-**50**. RS-66331 was reported to have pro-cognitive effects in a rat spatial memory test.[48,49] Administration of this compound reversed the atropine induced performance deficit in the rat as measured by performance in the Morris water maze. Evidence that 5-HT$_4$ receptor agonist activity was responsible for the procognitive activity of RS-66331 was provided from reversal of the effect by the selective 5-HT$_4$ antagonist RS-67532 (compound **101**, Table 1.11).

In another study of zacopride related structures, side chain modifications of the high affinity 5-HT$_3$ receptor antagonist/5-HT$_4$ receptor agonist ADR-932 (**52**) led to the discovery of the potent and selective 5-HT$_4$ receptor agonists (Table 1.6).[50] An initial structure-activity study showed that the 4-amino and 5-chloro substituents were required for binding to the 5-HT$_4$ receptor and that the ester (**53**) corresponding to ADR-932 was a potent 5-HT$_4$ receptor agonist (pEC$_{50}$ 8.0). However, as with ADR-932, the ester had a much higher affinity for the 5-HT$_3$ receptor (pK$_i$ 9.7). Much greater selectivity for the 5-HT$_4$ receptor was achieved by replacement of the quinuclidinyl side chain of ester **53** with the more flexible 4-(piperidinyl)methyl moiety which afforded the potent 5-HT$_4$ receptor agonist **54**. As noted in earlier series (e.g., related to ML 10302), esters were consistently higher affinity 5-HT$_4$ receptor ligands than their amide counterparts. However, the 1-piperidinylpropyl amide **59** was identified as reasonably potent 5-HT$_4$ receptor partial agonist with good selectivity over the 5-HT$_3$ receptor.

Examples of benzamides in which the *ortho*-methoxy group was replaced with heterocyclic rings capable of forming an intramolecular hydrogen bond include the zacopride and renzapride related compounds **60-63**.[51] Although these compounds demonstrated 5-HT$_4$ agonist properties in the field stimulated guinea pig ileum, they were also found to have high affinity for the 5-HT$_3$ receptor.

Other benzamides that have been claimed to have 5-HT$_4$ agonist properties are listed in Table 1.7, although the putative agonist activity of several of these was based on in vivo gastrointestinal prokinetic activity rather than receptor binding or functional assays. Of interest from an SAR point of view is the replacement of the usual *ortho*-methoxy group in the benzamide ring with a larger alkyl group in two of these putative agonists (LAS Z-019 and AS-4370), as it has been reported that substitution of ethoxy for methoxy in SC-53116 (**43**) resulted in a dramatic decrease in agonist activity.[9] In addition, a number of benzamides have appeared in the patent literature with claims of 5-HT$_4$ agonist activity; however, as these tend to repeat chemical themes already described, and are in general presented without substantiating data, they will not be dealt with in this chapter.

Aryl Ketones

As for the amides just described, aryl ketones related to the ester 5-HT$_4$ receptor ligands were also prepared in an attempt to overcome the metabolic lability of the latter compounds. Whereas bioisoteric replacement of the ester linkage of ML 10302 (**16**, Table 1.1) with a ketone resulted in a dramatic reduction in 5-HT$_4$ receptor agonist activity (as in ketone **34**),[33,36] similar modification of antagonist RS-23597 (**9**) had the interesting effect of producing a compound (**64**, RS-17017) with similar affinity for the 5-HT$_4$ receptor but with partial agonist properties (Table 1.8).[33] Higher affinity agonists were found in a related series of 3-(4-piperidinyl)propiophenones (**65-73**, Table 1.8).[33] Increasing the size of the alkyl group on the piperidine nitrogen led to a progressive increase in agonist activity, with maximal activity associated with an *n*-butyl (**70**) or 2-(methanesulfonamido)ethyl (**72**) substituent. From the disparate polarity of these two groups, it was suggested that size, rather than the polarity, was the major determinant of agonist activity. Larger substituents were detrimental to agonist activity.

The pharmacology of compounds **70** (RS-67333) and **72** (RS-67506) was assessed in greater detail in vitro and in vivo and these studies confirmed potent 5-HT$_4$ agonist properties.[57] Furthermore, both RS-67333 and RS-67506 were found to have good selectivity against other receptors of interest, including the 5-HT$_3$ receptor (5-HT$_3$ pK_i: RS-67333, 6.4; RS-67506, 5.6). In the rat Morris water maze model of spatial learning and memory, RS-67333 was found to be active when administered intraperitoneally; however, RS-67506 was inactive in this model.[23] It was suggested that this reflected the enhanced ability of the more hydrophobic RS-67333 to penetrate the CNS, relative to the more hydrophilic RS-67506. As previously observed for RS-66331,[48] the activity of RS-67333 in the water maze was abolished by pretreatment with the 5-HT$_4$ antagonist RS-67532.

RS-17017 (**64**) also demonstrated pro-cognitive activity in animal models including the rat Morris water maze and a primate delayed match-to-sample protocol in which the compound was orally active (RM Eglen, unpublished results).

Indoles

A molecular modeling study to map the 5-HT$_4$ receptor agonist pharmacophore was carried out using the commercially available package SYBYL (Tripos).[58] Low energy conformations of zacopride (**4**) were calculated and used to generate distance maps representing the spatial relationship between the presumed key pharmacophoric elements: the aromatic ring, the basic nitrogen, and the hydrogen-bond donor-acceptor function (methoxy group). Low energy conformations of 5-HT and a distance map were similarly determined, and an overlay with the zacopride map was performed taking into account low energy conformations and vector alignments of the basic nitrogen lone pair. The proposed binding orientation of zacopride in relation to 5-HT is represented in Figure 1.4. Key hypotheses generated from this modeling study were: 1) the indole NH and the amino substituent of zacopride hydrogen bond to the same acceptor; 2) the basic nitrogen of zacopride, which is ca. 2.0 Å farther from the center of the aromatic locus than in 5-HT, binds to a different mesomeric form of the receptor carboxylate; 3) zacopride does not contain functionality to mimic the 5-OH interaction of 5-HT, which accounts for the low affinity of zacopride; and 4) a lipophilic binding pocket exists near the basic nitrogen binding site. It is of interest to note that the proposed binding orientation for the benzamides at the 5-HT$_4$ agonist recognition site was different than that proposed for the binding of SC-53116 (Fig. 1.3).[42]

The 5-hydroxyindole carbazimidamide **74** (Fig. 1.4) met all of the above hypothetical criteria for binding to the 5-HT$_4$ agonist site and was, in fact, found to be a high affinity agonist (Table 1.9).[58] This compound then served as a useful scaffold to probe the requirements for binding within this structural class.[59] The higher affinity of the *N*-pentyl compound **75** and the *N*-phenylethyl derivatives **76** and **77** was consistent with the postulated lipophilic binding site. The importance of the 5-hydroxy group to agonist activity was confirmed by the low activity of the 5-methoxy compound **79**; however, the activity of the 5-allyloxy derivative **80** implied the existence of another small lipophilic binding region. Substitution of heterocyclic rings in the aminoguanidine moiety led to decreased activity, although imidazole **81** retained significant agonist affinity and efficacy. Replacement of the indole moiety with other aromatic rings gave a dramatic decrease in activity with the exception of the fused pyridone **82**.

In terms of selectivity over other serotonin receptors, ligand binding studies indicated that high affinity agonists **74** and **75** were 40- and 80-fold selective over the 5-HT_{1A} receptor and 25- and 20-fold selective over the 5-HT_{2C} receptor, respectively.[59] Neither compound displayed significant affinity for the 5-HT_3 receptor. Although these profiles are indicative of moderate selectivity, compounds **74** and **75** would appear to be the highest affinity, most potent agonists yet reported.

From this series of indole carbazimidamides the 5-methoxy derivative **79** (SDZ HTF 919) was chosen for evaluation as a gastrointestinal prokinetic agent. This partial agonist was active in vivo in animal models of large and small bowel transit and was advanced to human clinical trial.[60] The drug appeared to be tolerated upon twice daily oral administration to normal subjects, and the observation of increased colonic transit time as a surrogate measurement of prokinetic activity was felt to warrant further investigation in patients. The oral bioavailability of SDZ HTF 919 was reported to be 11% and a half-life of ca. 1.5 hours was observed.[61]

Other Compounds Reported to be 5-HT₄ Receptor Agonists

The tropine ester of benzofuran-3-carboxylic acid (**83**) was found to have antinociceptive activity in the mouse hot-plate test.[62] It was claimed that this activity resulted from a central cholinergic mechanism involving stimulation of central 5-HT_4 receptors on the basis that the analgesic effect was reversed by SDZ 205-557 (**7**). No further characterization of the purported 5-HT_4 agonist activity of **83** was reported. The gastrointestinal prokinetic agent VB20B7 (**84**) was claimed to be a weak 5-HT_3 receptor antagonist with 5-HT_4 receptor agonist properties.[63] In binding studies, this compound had essentially no affinity ($pK_i < 5$) for the 5-HT_4 receptor in rat cerebral cortex labeled with [³H]GR 113808. In the rat esophagus, VB20B7 reversed the contractile effect of carbachol with a pA_2 of 6.58 (compared to metoclopramide with a pA_2 of 6.98). VB20B7 also increased the twitch response in the electrically-stimulated guinea pig ileum at a concentration of 50 μM, an effect that was partially inhibited by tropisetron. Thus it was claimed that VB20B7 was a peripherally acting 5-HT$_4$ receptor agonist which was devoid of affinity for central 5-HT$_4$ receptors. This claim must, however, be regarded as speculative on the basis that there is no evidence for differences in central and peripheral 5-HT$_4$ receptors (following chapter) and that the weak activity observed in the esophagus and ileum could be accounted for by muscarinic antagonism.

5-HT$_4$ Receptor Antagonists

Benzoates and Benzamides and Related Systems

The benzoate ester SB 204070 (11) represented a major milestone in the development of potent and selective 5-HT$_4$ receptor antagonists.[28,64] The structure-activity relationships used to derive this compound are shown in Table 1.10.[28] The (4-piperidinyl)methyl ester 85 was found to be a more potent 5-HT$_4$ receptor antagonist than the earlier ester SDZ-205557 (7) by over two orders of magnitude. In a series of simplified benzoates (87-89) it was noted that incorporation of the *ortho*-oxygen into a six-membered ring (88) resulted in optimum activity. Addition of the amino and chloro substituents to the aromatic ring of benzodioxane 88 afforded the highly potent antagonist SB 204070 (11). Consistent with the SAR of ester 5-HT$_4$ receptor agonists, the amides 86 and 90 were less active than the corresponding esters; however, amide 90 (SB 205800) was nonetheless shown to be a high affinity antagonist (pK$_i$ 8.9 vs. [^{125}I]SB 207710 in rat cortex) with oral activity in the dog Heidenhain pouch model (ED$_{50}$ 3 µg/kg p.o.)[65]

Radioligand binding studies indicated that SB 204070 was remarkably selective (>5000-fold) over a broad range of other receptor subtypes.[66] Further evaluation in the guinea pig distal colon revealed a pA$_2$ value of 10.8 and increasing concentrations of SB 204070 were found to depress the maximum response to 5-HT.[66] This "nonsurmountable" antagonism was hypothesized to result from the high affinity of SB 204070 which caused the antagonist to behave in a "pseudo-irreversible" manner. A similar nonsurmountable antagonism was observed in the rat esophagus. In the conscious dog, SB 204070 antagonized the contractile effect of i.v. 5-HT in the Heidenhain pouch preparation with an ID$_{50}$ value of 0.55 µg/kg, i.v. SB 204070 potently inhibited 5-HT-induced tachycardia in the anesthetized micropig upon i.v. administration (ID$_{50}$ 4.7 µg/kg), but was inactive when administered intraduodenally,[67] thus underscoring the limitation of ester ligands for in vivo studies.

The high affinity and selectivity of SB 204070 led to the development of the radioligand [^{125}I]SB 207710 (**91**) which was shown to bind with high specificity and affinity (pK$_i$ 9.2) to piglet hippocampal 5-HT$_4$ receptors.[15] This radioligand is now commercially available from Amersham.

A series of imidazopyridines were evaluated in an attempt to find metabolically stable 5-HT$_4$ receptor antagonists amongst benzamide systems related to SC-53116 (**43**).[42] It was reasoned that because of the geometry of the imidazopyridine ring system, intramolecular hydrogen-bonding would be reduced relative to the benzamide system in which strong hydrogen-bonding exists between the amide NH and the *ortho*-methoxy group. Thus the amide side chain in the imidazopyridine system might assume an orientation closer to that in the benzoates (which lack the intramolecular hydrogen-bond), thereby resulting in increased antagonist properties. This hypothesis was borne out as the pyrrolizidine amide **92** (SC-53606) was found to be a moderately potent antagonist with a pA$_2$ of 7.9 in the rat esophagus.[68] Interestingly, the corresponding ester **93** was ca. 10-fold less active (pA$_2$ 7.0), one of the few reported examples in which an amide was more active than its ester counterpart. On the basis of ligand binding, SC-53606 was 185-fold selective for the 5-HT$_4$ over the 5-HT$_3$ receptor. Many other benzoate/benzamide derived bi- and tricyclic systems have been reported in the patent literature to possess 5-HT$_4$ receptor antagonist activity, and these have been tabulated in a previous review.[10]

Compound **94** is an example of a benzamide containing a piperazinyl side chain which been reported to be a 5-HT$_4$ receptor antagonist with a pK$_b$ of 7.4 in the rat esophagus.[69]

Aryl Ketones

Two separate structural modifications of the aryl ketone 5-HT$_4$ receptor agonist RS-67506 (**72**) resulted in the potent antagonists **95** (RS-39604)[70] and **96** (RS-100235).[71] Preliminary SAR work based on the agonist RS-17017 (**64**) indicated that replacement of the *ortho*-methoxy group with benzyloxy conferred antagonist properties in the series (Table 1.11).[70] The dimethoxybenzyloxy compound **101** (RS-67532) was subsequently used as a tool in studies of CNS effects of 5-HT$_4$ receptor agonists.[23,48] Substitution at the 4-position of the piperidinyl moiety of antagonist **101** was tolerated and resulted in a slight increase in antagonist potency (compound **111**). Incorporation of these structural features into the RS-67506 series afforded higher affinity antagonists, as exemplified by RS-39604 (**95**, Table 1.12), an antagonist with sub-nanomolar affinity for the 5-HT$_4$ receptor. In the anesthetized, vagotomized micropig RS-39604 inhibited 5-HT-induced tachycardia with ID$_{50}$ values of 4.7 µg/kg and 254 µg/kg by i.v. and intraduodenal routes of administration, respectively.[67] The compound also inhibited 5-hydroxytryptophan-induced diarrhea in the conscious mouse with ID$_{50}$ values of 81 µg/kg, i.p., and 1.1 mg/kg, p.o.

Combination of structural features from SB 204070 (**11**) and the aryl ketone series produced the potent 5-HT$_4$ antagonists shown in Table 1.13.[71] Introduction of the lipophilic 3,4-(dimethoxy)phenylpropyl substituent at the piperidine nitrogen gave RS-100235 (**96**), which as previously observed with SB 204070,[28] was a nonsurmountable antagonist in the rat esophagus. In accord with the in vitro functional data, RS-100235 was a significantly more potent antagonist of 5-HT-induced tachycardia in the micropig (ID$_{50}$ 0.55 µg/kg, i.v., and 1.5 µg/kg, i.d.) than progenitor RS-39604. A receptor binding profile of RS-100235 showed this ligand to have excellent selectivity over other receptor subtypes of interest.

On the basis of its pharmacokinetic profile in animals, compound **119** (RS-100302) was chosen from this series for clinical evaluation. RS-100302 was well tolerated in normal subjects with excellent oral bioavailability and a prolonged plasma half-life (RM Eglen, unpublished results).

Indoles and Indazoles

The indole ester GR 113808 (**10**) provided an early structural lead in the search for selective 5-HT$_4$ receptor antagonists and an important tool for the pharmacological characterization of the 5-HT$_4$ receptor.[27] GR 113808 was an antagonist of 5-HT-induced contraction in the guinea pig ascending colon with a pA$_2$ value of 9.2, and similarly antagonized 5-HT-induced relaxation of the carbachol-contracted rat esophagus with a pA$_2$ of 9.3. The compound was also a potent antagonist of 5-methoxytryptamine (5-MeOT)-induced tachycardia in the anesthetized piglet. GR 113808 was shown to be over 1000 times selective over the 5-HT$_3$ receptor with similarly low affinity for other receptors. [^{3}H]GR 113808 (labeled in the N-Me group) provided an invaluable and commercially available (Amersham) radioligand for 5-HT$_4$ receptor binding studies.[14]

The more highly elaborated analog **122** (GR 125487) was subsequently reported to be a more potent $5\text{-}HT_4$ antagonist than GR 113808 with pA_2 values of 10.0 in both the rat esophagus and guinea pig proximal colon.[72] Determination of the $5\text{-}HT_4$ receptor affinity of GR 125487 by radioligand binding in guinea pig striatal membranes furnished a pK_i value of 10.4. In the anesthetized piglet GR125487 was more potent than GR 113808 in inhibiting 5-MeO-5HT-induced tachycardia (DR_{10} 0.29 µg/kg, i.v.).[73] In another in vivo model, GR 125487 was found to inhibit the metoclopramide-induced increase in gastric emptying in the rat with an ID_{50} of 0.78 µg/kg when administered subcutaneously. Data from a patent application indicated that a ketone analog of GR 113808 (*n*-butyl side chain) was a weak $5\text{-}HT_4$ receptor antagonist (pIC_{50} 6.8 in guinea pig colon).[74]

SB 203186 (**123**) was another indole ester with $5\text{-}HT_4$ antagonist properties that provided early evidence for the involvement of $5\text{-}HT_4$ receptors in various physiological responses, most notably in cardiac tissues from several species including man.[75] The pA_2 values for $5\text{-}HT_4$ antagonist activity of SB 203186 in human atrium and piglet atrium were 8.7 and 8.3, respectively. The effect of conformational constraint of the side chain of indole esters related to SB 203186 was evaluated in a series of azabicyclic derivatives (Table 1.14).[76] The results indicated that there was an optimum distance between the ester group and the basic nitrogen, as typified by examples **125**, **126**, **129**, **131** and **132**. However, steric factors were also implicated as several derivatives with a similar spacer, e.g., **127**, were much less active. A separate study of related azabicyclic indole esters identified the pyrrolizidines **134** (pA_2 8.4, rat esophagus) and **135** (SC-56184, pA_2 9.2) as high affinity $5\text{-}HT_4$ antagonists.[42]

Although many of the aforementioned indole esters represented high affinity $5\text{-}HT_4$ antagonists that were clearly useful for in vitro evaluation, their limited bioavailability was expected to severely limit in vivo utility. Thus another milestone in the $5\text{-}HT_4$ receptor field was the discovery of the orally active indole amide **143** (SB 207266).[77] Following the SAR work that led to SB 204070 (**11**), it was found that incorporation of the oxygen adjacent to an ester or amide side chain into a six-membered ring led to a significant enhancement of antagonist potency.[78] Thus the oxazino-[3,2-a]indole ester **142** and, more importantly, amide **143** were more active than the *ortho*-methoxy progenitors **138** and **139** (Table 1.15).[77] The SAR in this series were consistent with those in the SB 204070 work, e.g., the six-membered oxo-containing ring derivative (143) was at least 10-fold more active than the 5- and 7-membered ring congeners (141 and 145). From examples 142 and 146 it would also appear that the (4-piperidinyl)methyl side chain conferred increased 5-HT$_4$ antagonist properties relative to the (2-quinolizinyl)methyl side chain.

Detailed pharmacological evaluation of SB 207266 revealed excellent selectivity over other receptors and functional pharmacology expected for a potent 5-HT$_4$ receptor antagonist.[79] A pA$_2$ value of 10.6 was obtained in the guinea pig distal colon and nonsurmountable antagonism was observed in this preparation. In the dog Heidenhain pouch model, SB 207266 produced antagonism of 5-HT-induced contractions with ID$_{50}$ values of 1.3 µg/kg, i.v. and 9.6 µg/kg, p.o., and the effect observed after oral dosing appeared to be of significant duration. SB 207266 was reported to be in Phase II clinical trials for treatment of irritable bowel syndrome and was reportedly well absorbed and tolerated in normal volunteers and patients.[80] Another report described anxiolytic-like activity for SB 207266, as well as SB 204070, in 2 rat models: an elevated x-maze and a social interaction test.[81]

A limited number of examples of the use of indazole as an indole bioisostere have been reported in the 5-HT$_4$ receptor antagonist literature. The indazole analog 147 was reported to be slightly more potent and selective (vs. the 5-HT$_3$ receptor) than its indole counterpart tropisetron.[82] On the other hand, indazole 148 was found to have 4-fold less 5-HT$_4$ receptor antagonist activity than indole 135.[42] The indazole amide 149 (LY353433) was shown to be a 5-HT$_4$ antagonist in vitro in the rat esophagus, although an affinity constant was not reported.[83] After oral administration in the rat, LY353433 inhibited 5-HT$_4$ receptor-mediated relaxation of the esophagus ex vivo with an ID$_{50}$ of 0.1 mg/kg. Selectivity data against other 5-HT receptors were not reported. The indazole amide 150 (N-3389) was reported to have the same 5-HT$_4$ antagonist potency as tropisetron in the guinea pig ileum; however, this compound was also reported to have a pK$_i$ of 8.7 at the 5-HT$_3$ receptor.[84]

Indol-3-yl carbazimidates with 5-HT$_4$ antagonist properties were produced through minor structural modifications of the potent agonist 74.[85] It was speculated that substitution of a small alkyl group at position-1 or -7 of the indole nucleus of 74 would displace the ligand from the agonist binding site without significantly reducing receptor affinity. In the event, ligands 151-155 (Table 1.16) were indeed found to be 5-HT$_4$ receptor antagonists with reasonable affinity. As had been noted in the agonist series (Table 1.9), the 5-hydroxy group was important for receptor binding as methylation of hydroxy compound **152** led to a significant decrease in activity (**153**). Binding studies indicated that whereas compounds **151** and **154** had relatively high affinity for the 5-HT$_{1D}$ receptor (pK$_i$ of 7.8 and 7.9, respectively), the *N*-methyl analog **152** had much lower affinity for this receptor (pK$_i$ 5.4). Thus compound **152** had the best overall selectivity profile being ca. 25-fold selective over the 5-HT$_{2C}$ receptor.

Carbamate Derivatives

Aryl carbamates and ureas substituted with an oxadiazole or thiadiazole ring in the *ortho*-position were claimed in a patent to have 5-HT$_4$ antagonist properties. Data presented for selected examples in the oxadiazole series indicated that carbamates **156** and **160** were found to be potent 5-HT$_4$ receptor antagonists (Table 1.17).[86] On the basis of the two examples **157** and **158**, ring substitution appeared to be detrimental to affinity, as did replacement of the carbamate linkage with urea (compound **161**).

More extensive SAR were published on another series of aryl carbamates derived by replacement of the ester linkage of agonist ML 10302 (**16**).[87] The highest affinity ligands from this study (Table 1.18) were devoid of 5-HT$_4$ receptor agonist activity as determined by functional testing in the guinea pig ileum. The aromatic amino and chloro substituents were not required for optimum binding which represented a departure from most of the benzoates and benzamides described in this chapter. Disubstitution with methyl groups in positions-3 and -5 of the piperidine was associated with an increase in 5-HT$_4$ receptor affinity, although this did not always translate to an increase in functional antagonist activity (e.g., **165** vs. **169**). Other types of aryl carbamates claimed in the patent literature as 5-HT$_4$ receptor antagonists include the tetrahydrocarbazole **170**[88] and the benzimidazolone **171** (pK$_i$ 8.9 in pig striatum).[89]

Bioisosteric Replacement of Amide and Ester Linkages

Several patent claims to 5-HT$_4$ receptor ligands in which an amide or ester linkage was replaced by a heterocyclic ring have been published. The oxadiazolyl indole **172** and related compounds were claimed as 5-HT$_4$ antagonists (pIC$_{50}$ 7.3 for **172** in guinea pig colon).[90] Oxadiazolones of the types **173** and **174** were claimed to have affinity for 5-HT$_4$ receptors with no data given.[91,92]

Pharmacophore for 5-HT$_4$ Receptor Ligands

A pharmacophore model for 5-HT$_4$ receptor antagonists is shown in Figure 1.5. Key features of the pharmacophore are an aromatic ring to which a carbonyl group is attached and a basic nitrogen in the appended side chain. The distance relationships between these elements were derived from X-ray crystallographic data for SB 204070 and SB 205800 and computer assisted conformational analysis.[10] Another key feature of the pharmacophore is an oxygen atom adjacent to the carbonyl group as is in the benzodioxane and oxazinoindole (e.g., SB 207266) antagonists. Presumably both the carbonyl group and adjacent ring oxygen atom serve as hydrogen bond acceptors; the placement of the oxygen atom in a ring may facilitate the hydrogen bond interaction by presenting an optimum orientation to the receptor and/or by removing an unfavorable steric interaction. It has been suggested that the higher affinity of ester antagonists relative to amides results from conformational differences arising from rotations about bonds α and β (Fig. 1.6), although it was acknowledged that differences in hydrogen bond capabilities could also be involved.[10] The 5-HT$_4$ receptor appears to provide a large pocket accessible to basic nitrogen substituents (e.g., RS-100235), although added side chain bulk does not necessarily increase affinity.

Carbamate antagonists such as compound **156** share these key pharmacophoric elements with the proviso that the aromatic ring would superimpose with the benzene ring of indole antagonists. The heteroatoms in the oxadiazole ring of **156** could provide the hydrogen bond capability of the ring oxygen in SB 207266 (Fig. 1.6).

The close structural similarity between 5-HT$_4$ agonists and antagonists implies that the pharmacophoric elements for binding to the 5-HT$_4$ receptor are virtually identical for both classes of ligands. As is evident from numerous examples presented in this chapter, what would appear to be relatively subtle structural changes can result in interconversion of agonist and antagonist properties. For example, benzofuran **54** (Table 1.6) and benzodioxan SB 204070 (**11**) share all of the requisite elements for binding to the antagonist binding site (Fig. 1.5), yet ligand **54** is a high affinity full agonist and SB 204070 is a high affinity antagonist. Similarly, substitution of methyl for hydrogen in the indole carbazimidamide **74** (Table 1.9) resulted in conversion of a potent agonist into an antagonist (compound **151**, Table 1.16). Furthermore, substitution at the seemingly remote terminus of the side chain of antagonist **85** (Table 1.10) produced a ligand (**175**, RS-57639) with partial agonist properties in some preparations (e.g., pEC$_{50}$ 9.0, 50% efficacy of 5-HT in rat esophagus).[93] As the only selective 5-HT$_4$ receptor agonist radioligand yet described, [^{3}H]RS-57639 could provide a useful tool for the study of ligand interactions.[94]

Conclusion

Tremendous advances in the medicinal chemistry of 5-HT$_4$ receptor ligands were made in a relatively short time subsequent to initial reports of nonselective lead compounds. High affinity ester ligands were discovered which formed the basis for medicinal chemistry programs directed toward improved pharmacokinetic properties. The liabilities of these ester ligands was overcome and potent, orally bioavailable 5-HT$_4$ receptor agonists and antagonists were produced from different chemical series. Unlike the 5-HT$_3$ receptor antagonist field, from which the early leads for 5-HT$_4$ receptor ligands were forthcoming, antagonists of the 5-HT$_4$ receptor share close structural similarities with agonists of the same receptor and appear to bind in a strikingly similar manner. The elucidation of the subtle factors responsible for this intriguing phenomenon poses a major challenge to medicinal chemists and molecular biologists alike.

Structures and Figures

1 (Tropisetron)   2 (Cisapride)   3 (Renzapride)

4 (Zacopride)   5 (BRL 20627)   6 (Metoclopramide)

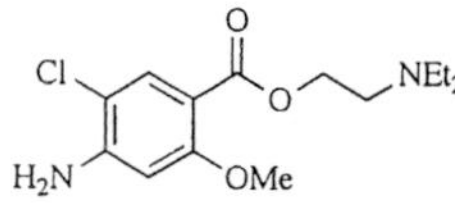

7 (SDZ 205-557)

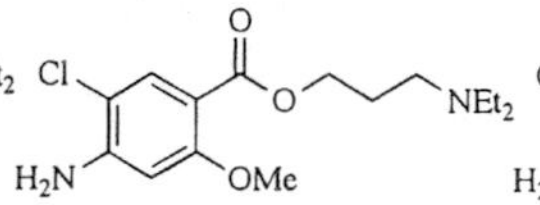

8 (LY 297524)

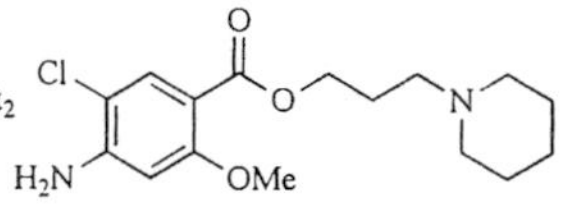

9 (RS-23597)

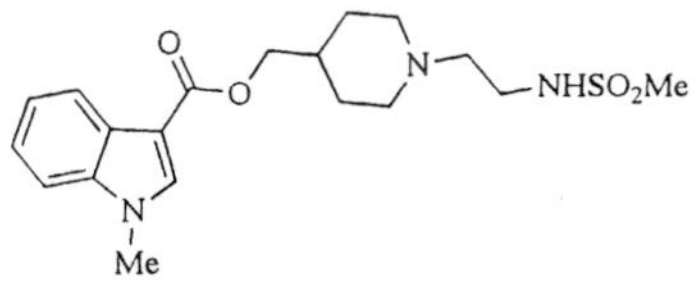

10 (GR 113808)

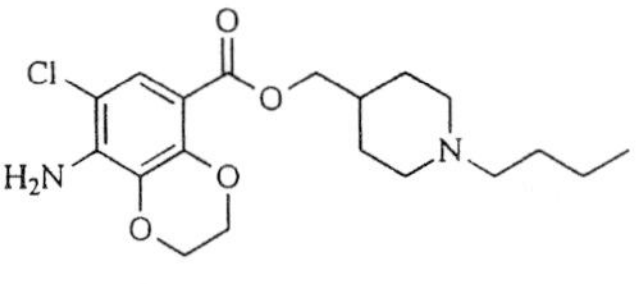

11 (SB 204070)

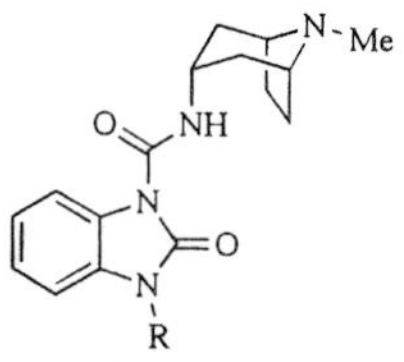

12 R = 2-Pr (BIMU 8)
13 R = H (DAU 6215)
14 R = Et (BIMU 1)

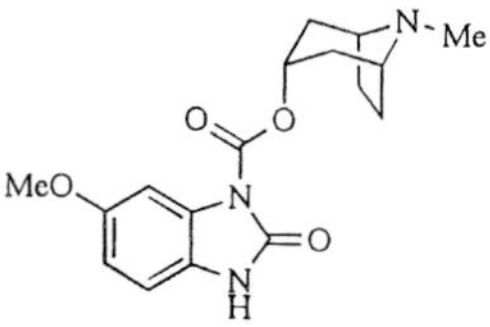

15 (DAU 6285)

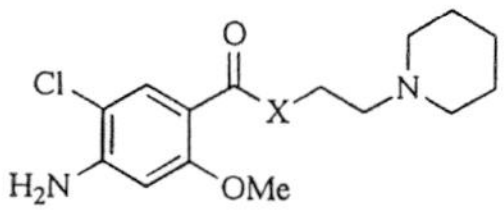

16 X = O (ML 10302)
33 X = NH
34 X = CH₂

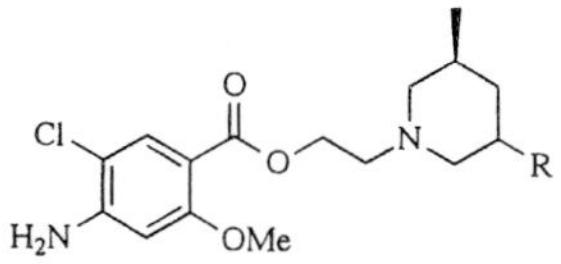

31 R = *cis* -Me
32 R = *trans*- Me

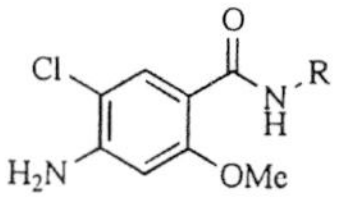

5 R =

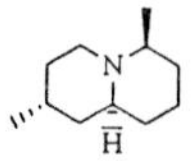

35 R =

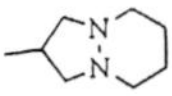

36 R =

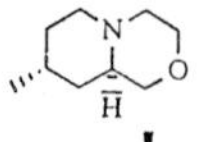

37 R =

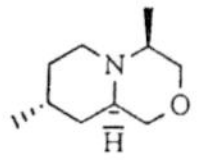

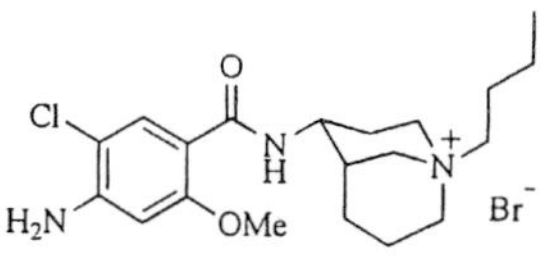

38 (SB 205149)

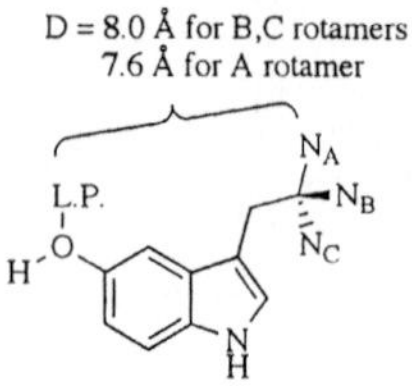
D = 8.0 Å for B,C rotamers
7.6 Å for A rotamer
L.P.
N_A
N_B
N_C

Figure 1.1 Proposed active conformer for 5-HT_4 agonism

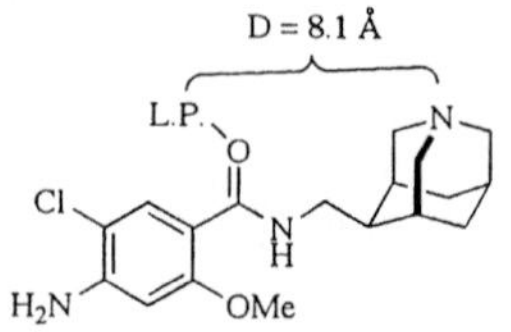
D = 8.1 Å
L.P.

Figure 1.2 Lone pair-nitrogen distance for aza-adamantane **39**

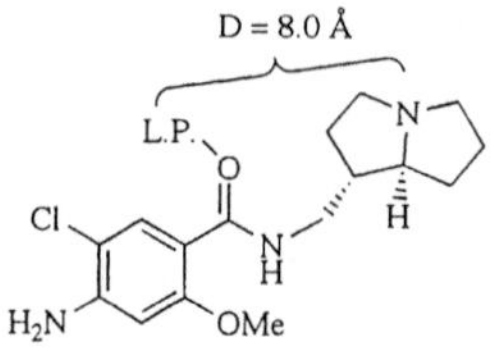
D = 8.0 Å
L.P.

Figure 1.3 Lone pair-nitrogen distance for pyrrolizine **43** (SC-53116)

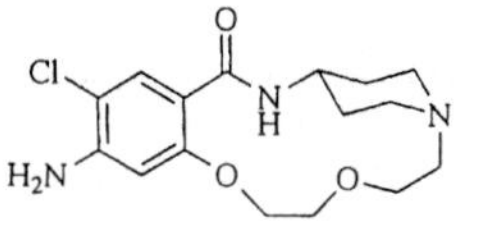

47

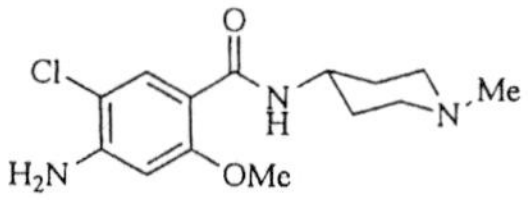

48

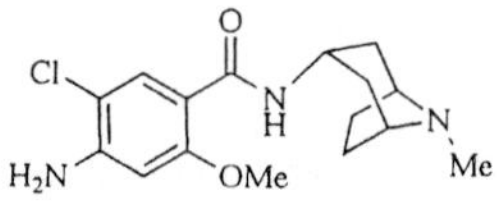

49

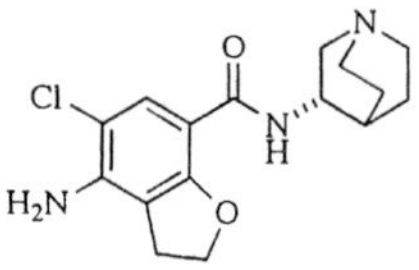

52 (ADR-932)

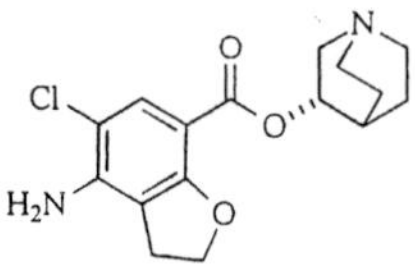

53

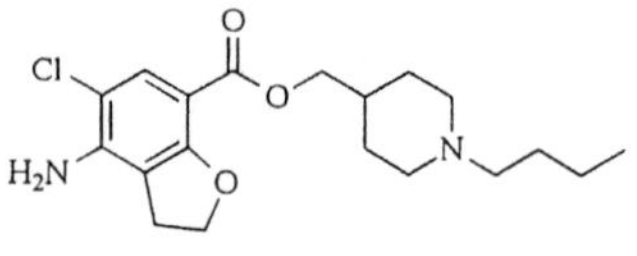

54 (FCE 29029)

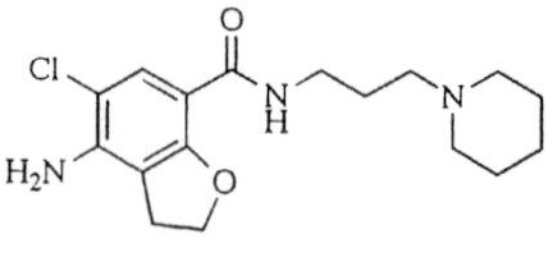

59 (FCE 29034)

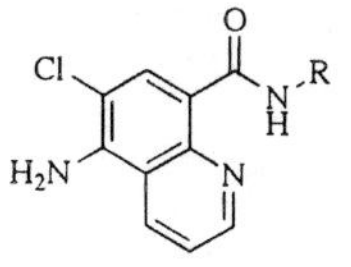

60, 61 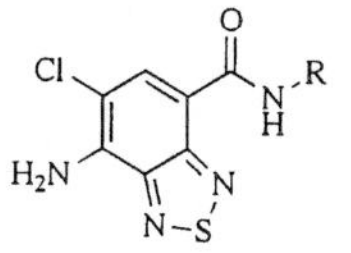   **62, 63**   **60** and **62** R =

61 and **63** R =

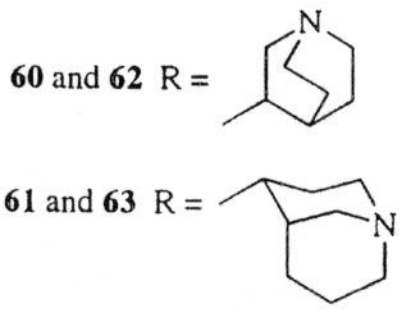

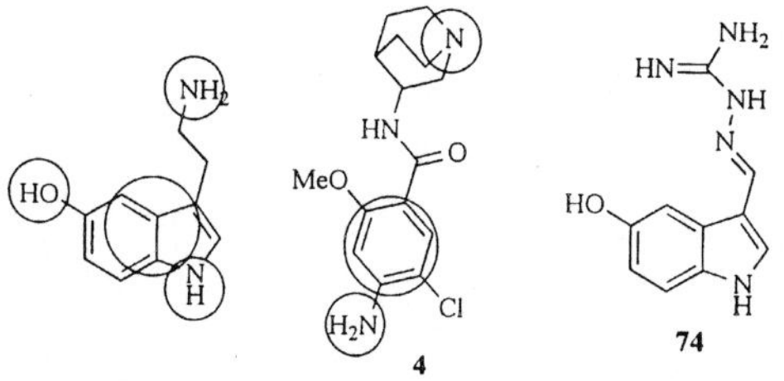

4   **74**

Figure 1.4 Important regions of interaction of 5-HT and zacopride (**4**) with the 5-HT₄ receptor. Structure of agonist **74** shown at right.

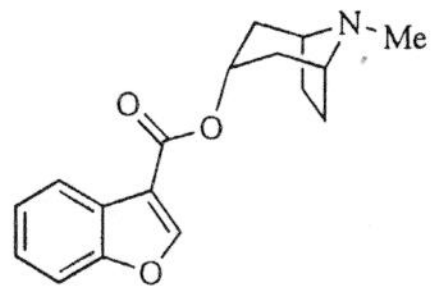

83 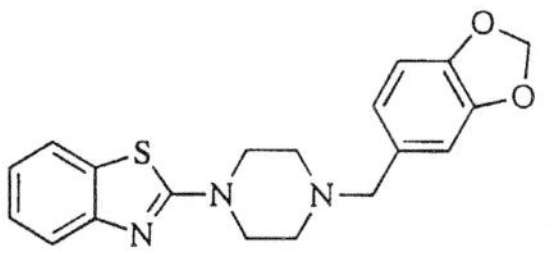   **84** (VB20B7)

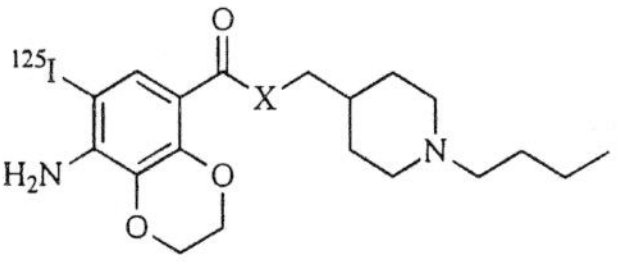

91 [¹²⁵I]SB 207710

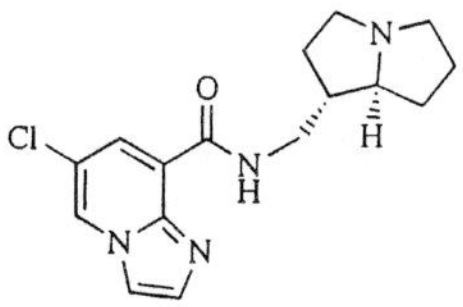

92 (SC -53606) 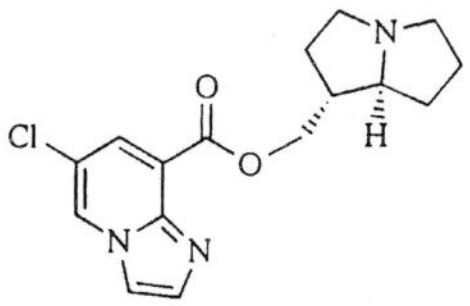   **93**

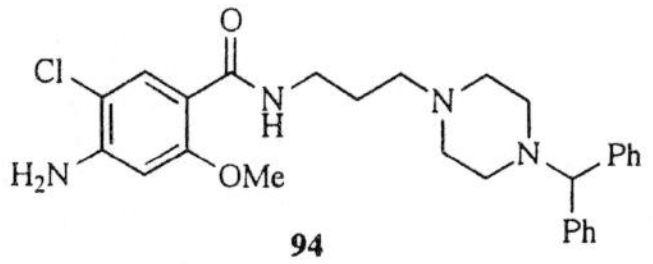

94

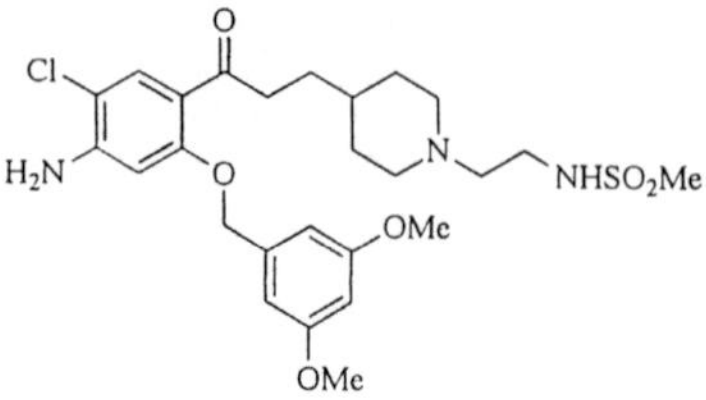

95 (RS-39604)

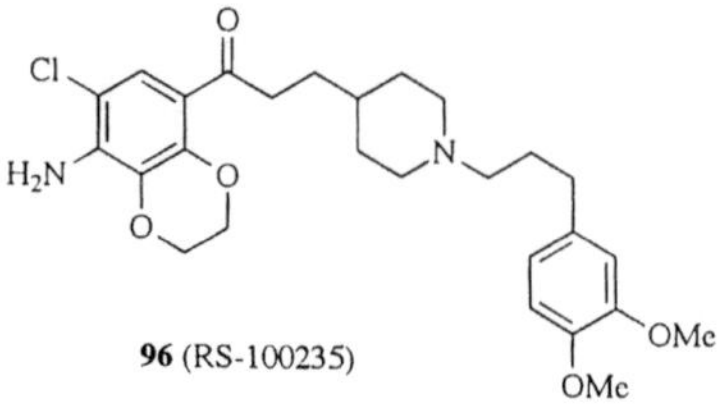

96 (RS-100235)

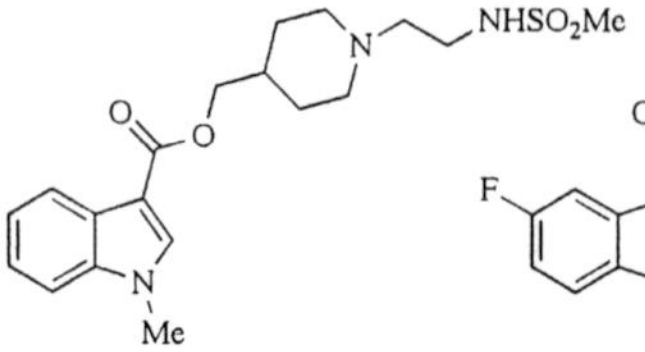

10 (GR 113808)

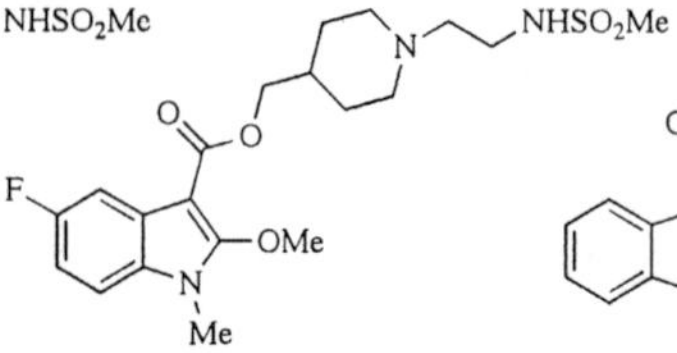

122 (GR 125487)

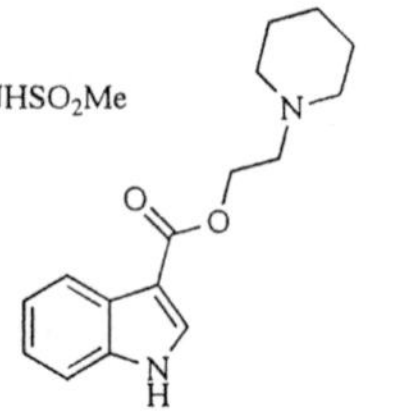

123 (SB 203186)

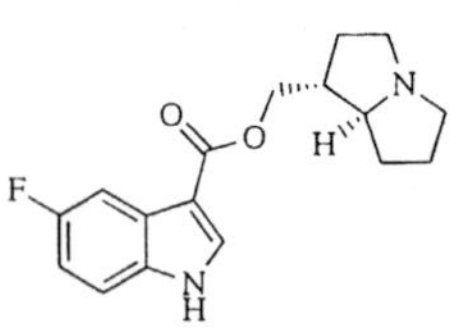

134

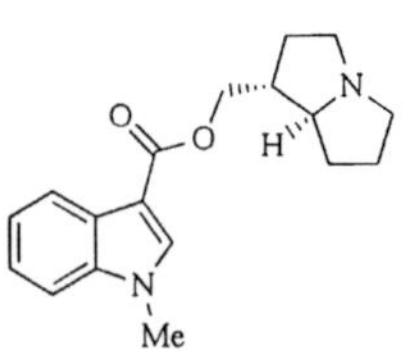

135 (SC-56184)

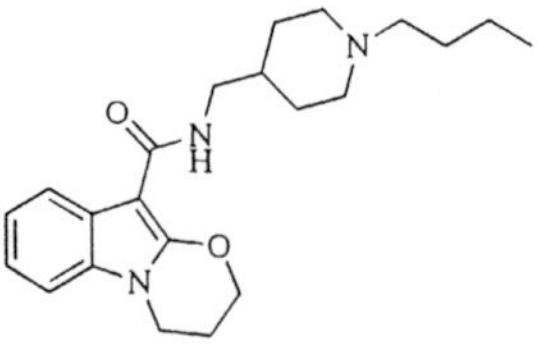

143 (SB 207266)

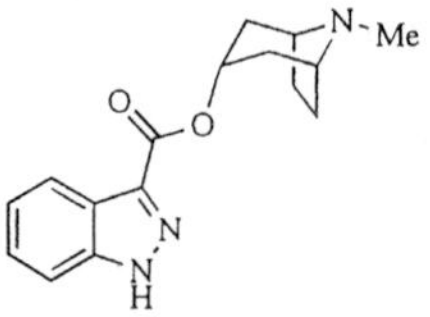

147

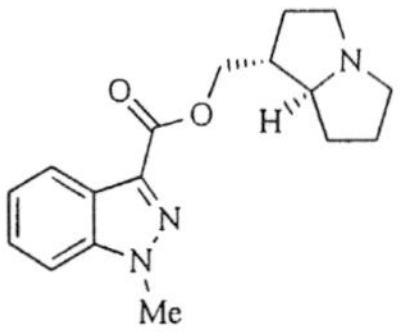

148

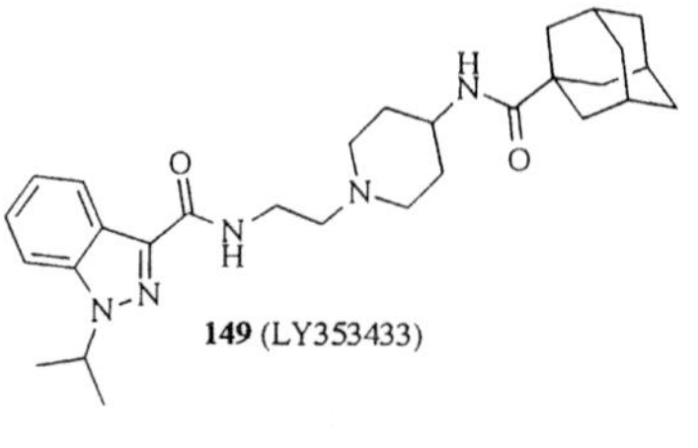

149 (LY353433)

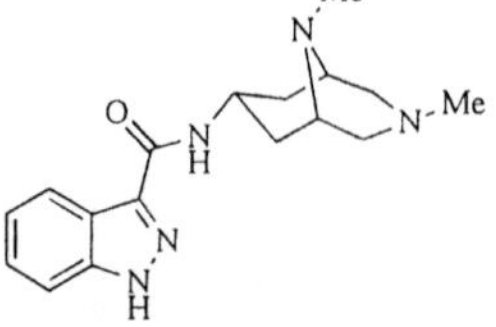

150 (NS-3389)

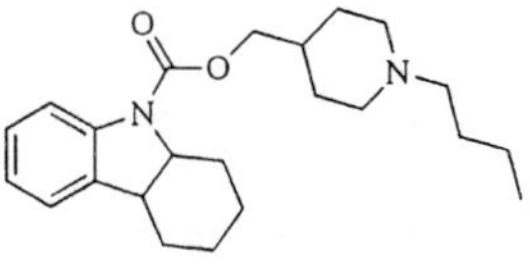

170

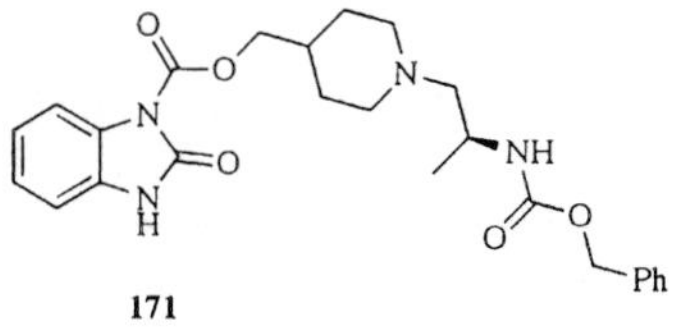

171

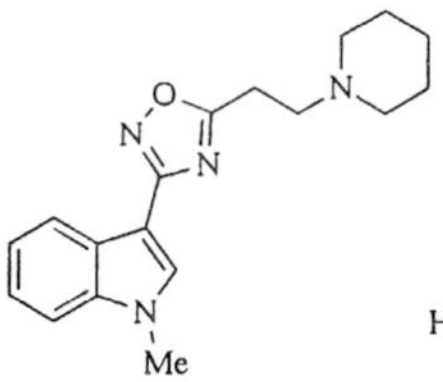

172

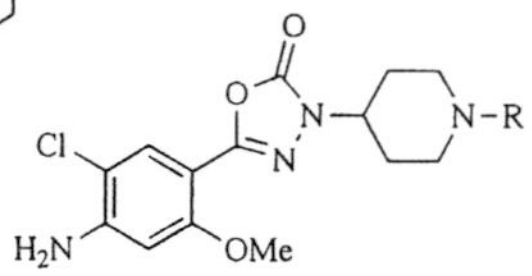

173

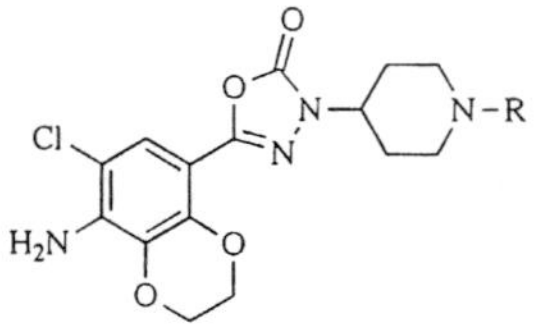

174

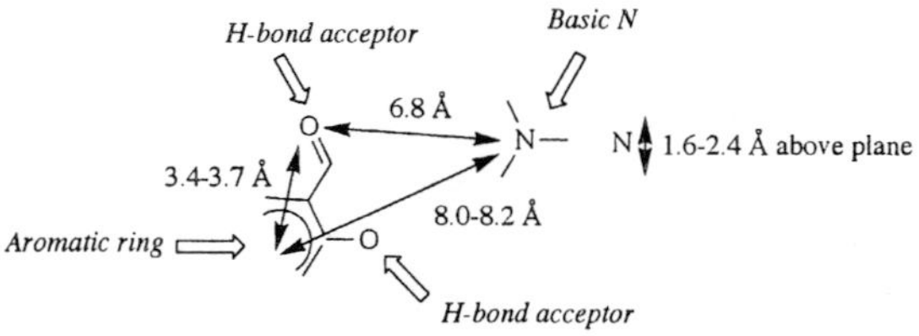

Figure 1.5. A proposed pharmacophore for 5-HT$_4$ receptor antagonists.

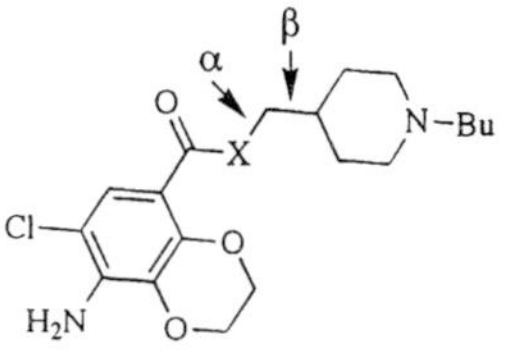

SB 204070 X = O

SB 205800 X = NH

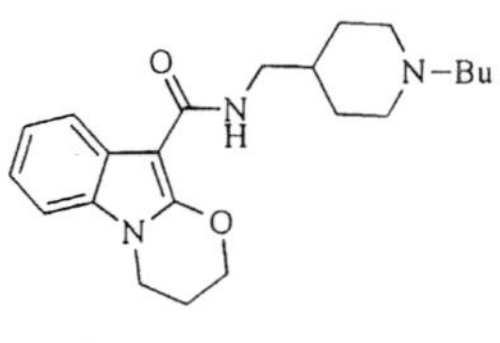

SB 207266

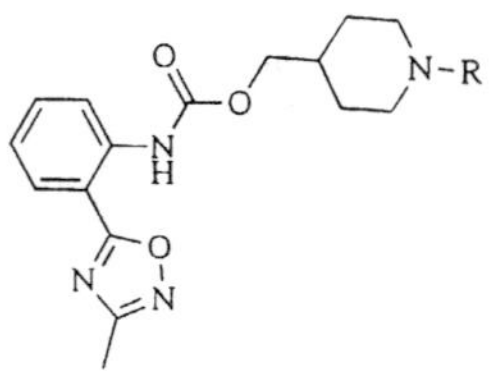

156 (R = CH$_2$CH$_2$NHSO$_2$Me)

Figure 1.6. High affinity 5-HT$_4$ receptor antagonists from ester, amide and carbamate series for comparison with pharmacophore model in Figure 1.5.

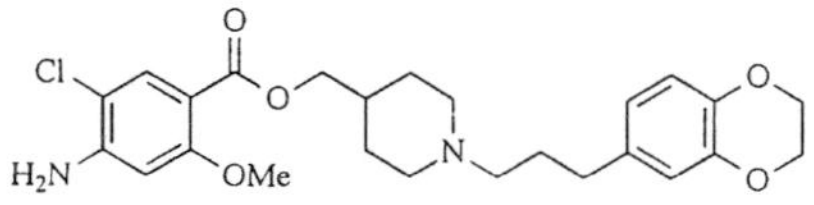

175 (RS-57639)

Tables

TABLE 1.1. 5-HT$_4$ Receptor Activity for Piperidinoethyl Benzoates[a]

compd.	R	Binding pK$_i$[b]	Agonist pEC$_{50}$[c]	% E[d]
16	H	9.0	8.4	80
17	2-Me	8.1	8.2	82
18	3-Me	9.0	7.8	54
19	4-Me	8.4	8.5	57
20	3-OH	8.3	8.3	64
21	4-OH	8.8	7.9	72
22	4-NHCOMe	9.0	7.9	59
23	4-CONH$_2$	8.8	7.7	65
24	4-Benzyl	8.1	inactive	—
5-HT		6.7	8.3	100
SDZ 205-557 (**7**)		8.3	inactive	—
GR 113808 (**10**)		9.8	inactive	—

[a]Reference 36. [b]Displacement of [^{3}H]GR 113808 from rat striatal membranes. [c]Electrically stimulated guinea pig ileum. [d]Maximum % stimulation relative to 5-HT in guinea pig ileum.

TABLE 1.2. 5-HT$_4$ receptor activity for aminoethyl benzoates[a]

compd.	NR$_1$R$_2$	Binding pK$_i$[b]	Agonist pEC$_{50}$[c]	% E[d]
25		8.3	7.4	50
17		9.0	8.4	80
26		9.0	8.4	59
27		8.0	7.4	63
28		8.3	8.1	61
29		8.2	8.3	78
30		8.2	7.8	76

[a]Reference 36. [b]Displacement of [^{3}H]GR 113808 from rat striatal membranes. [c]Electrically stimulated guinea pig ileum. [d]Maximum % efficacy relative to 5-HT in guinea pig ileum.

TABLE 1.3. Pharmacological data for azacyclic benzamides[a]

compd. (series)	R	5-HT$_4$ Receptor Activity		5-HT$_3$ Activity
		Agonist EC$_{20}$[b]	HP, LAD[c]	ID$_{50}$[d]
38 (exo)		47	0.05	45
39 (endo)		17	0.01	3.0
40 (endo)		0.4	0.005	30
3 (endo)		7	0.01	3.3
41 (exo)		87	0.05	>100
42 (endo)		–	0.01	3.4
43 (exo)		–	0.05	59
44		–	0.01	86
45 (endo)		–	0.01	19

[a]Reference 40. [b]Electrically stimulated guinea pig ileum, ng/mL. [c]Dog Heidenhain pouch, lowest active dose, mg/kg, iv. [d]Inhibition of Bezold-Jarish reflex in the rat, ug/kg, iv.

TABLE 1.4. Pharmacological data for azacyclic benzamides[a]

Compd.	R	Lone pair-N Distance, Å[b]	5-HT₄ Agonist Activity pEC_{50}[c]	5-HT₃ Binding pK_i[d]
39		8.1	7.1	–
40		7.3	6.3	6.5
41[e]		7.0	7.3	8.9
42[e]		7.8	6.8	–
43[e]		8.0	7.8	6.8
44[e]		8.0	6.5	7.2
45[f]		8.4	6.6	6.6
46[f]		7.0	6.8	–
5-HT		7.6-8.0	7.8	6.4

[a]References 9, 42. [b]See figure 1.1. [c]Relaxation of carbachol contracted rat esophageal muscularis mucosae. [d]Displacement of [³H]zacopride. [e]Pure enantiomer. [f]Racemate.

TABLE 1.5. Pharmacological data for naphthalimides[a]

compd.	R	5-HT$_4$ Receptor Activity			5-HT$_3$ Binding
		pK$_i$[b]	pEC$_{50}$[c]	% E[d]	pK$_i$[e]
(S)-50		7.6	7.9	80	8.0
(R)-50		6.5	<6	–	9.1
51	N-Me	–	7.5	90	7.4
(S)-zacopride ((S)-4)		6.9	7.2	90	9.6
(R)-zacopride ((R)-4)		6.2	6.3	90	8.5
5-HT		7.5	8.2	100	–

[a]References 46, 47. [b]Displacement of [³H]GR 113808 from guinea pig striatum. [c]Relaxation of carbachol contracted rat esophageal muscularis mucosae. [d]Maximum % relaxation relative to 5-HT. [e]Displacement of [³H]quipazine from rat cerebro cortical membranes.

TABLE 1.6. Pharmacological data dihydrobenzofurans[a]

compd.	X	5-HT$_4$ Receptor Activity			5-HT$_3$ Binding
		pK$_i$[b]	pEC$_{50}$[c]	% E[d]	pK$_i$[e]
54		9.9	8.7	100	6.6
55		8.7	8.3	59	7.2
56		9.0	8.7	71	6.9
57		7.4	7.5	65	6.4
58		8.3	8.1	70	5.9
59		8.0	7.7	60	<5
ADR-932 (**52**)		7.8	7.7	–	10.0
5-HT		7.4	7.7	100	1

[a]Reference 50. [b]Displacement of [^{3}H]GR 113808 from rat striatum. [c]Relaxation of carbachol contracted rat esophageal muscularis mucosae. [d]Maximum % relaxation relative to 5-HT. [e]Displacement of [^{3}H]BRL 43694 from rat entorhinal cortex.

TABLE 1.7. *Other benzamides reported to be 5-HT₄ receptor agonists*

compd.	R	X	Reported activity	Ref.
LAS Z-019	*n*-Pr		pEC$_{50}$ 7.2 (56% E)[a] gastro prokinetic[b]	52
R 76186	Me		contraction of guinea pig colon[c]	53
—	Me		pEC$_{50}$ 9.0[d]	54
AS-4370 (mosapride)	Et		gastro prokinetic[e]	55
TKS 159	Me		enhancement of e.s. evoked contraction of g.p. stomach	56

[a]Rat esophagus. [b]In vivo in the dog. [c]Contraction of isolated guinea pig colon ascendens. [d]Contraction of guinea pig ascending colon. [e]In several species including man.

TABLE 1.8. 5-HT$_4$ receptor agonist activity for aryl ketones[a]

compd.	R	pEC$_{50}$[b]	% E[c]
64		7.7	50
65	H	6.9	52
66	Me	7.4	53
67	Et	7.8	61
68	Pr-*n*	7.9	66
69	allyl	8.0	30
70	Bu-*n*	8.7	60
71	hexyl-*n*	7.2	55
72	(CH$_2$)$_2$NHSO$_2$Me	8.8	50
73	(CH$_2$)$_3$NHSO$_2$Me	antagonist	—
ML 10302 (**16**)		8.0	50
5-HT		8.2	100

[a]Reference 33. [b]Relaxation of the carbachol contracted rat esophageal muscularis mucosae.
[c]Maximum % relaxation relative to 5-HT.

TABLE 1.9 5-HT$_4$ receptor agonist activity for carbazimidamides[a]

compd.	R$_1$	R$_2$	pD$_2$[b]	% E[c]
74	OH	H	8.8	150
75	OH	pentyl	9.3	90
76	OH	CH$_2$CH$_2$Ph	9.1	100
77	OH	CH$_2$CH$_2$-3,4-Cl$_2$C$_6$H$_3$	>10	10
78	H	pentyl	<6	10
79	OMe	pentyl	6.9	20
80	OCH$_2$CH=CMe$_2$	pentyl	8.1	220
81			8.2	100
82			7.8	90
5-HT			8.4	100

[a]References 58, 59. [b]Enhancement of the twitch response in the electrically stimulated guinea pig ileum.
[c]Maximum % efficacy relative to 5-HT.

TABLE 1.10. SAR of the 5-HT$_4$ receptor antagonist SB 204070[a]

compd.			pIC$_{50}$[b] (pA$_2$)[b]
85	X = O		9.0
86	X = NH		8.0
87	n = 1		7.2
88	n = 2		8.2
89	n = 3		7.3
11 (SB 204070)	X = O		10.1 (10.8)
90 (SB 205800)	X = NH		— (10.0)
tropisetron (**1**)			5.5
SDZ 205-557 (**7**)			6.6

[a]References 28, 65. [b]Inhibition of 5-HT evoked contractions in the guinea pig distal colon

TABLE 1.11. SAR of aryl ketone 5-HT$_4$ receptor antagonists [x]

compd.	R	pK$_b$ [b]	compd.	X	pK$_b$ [b]
64	Me	—[c]	101	H (RS-67532)	8.5
97	Et	<7	103	Me	8.3
98	Pr-*n*	—[c]	104	Pr-*n*	7.7
99	CH$_2$—⬡—OMe	8.4	105	Ph	7.7
100	CH$_2$—⬡(OMe)—OMe	8.5	106	OH	8.3
			107	OMe	8.4
101	CH$_2$—⬡(OMe)(OMe)	8.5	108	CONH$_2$	8.4
			109	NHCONH$_2$	8.5
			110	NHSO$_2$Me	8.4
102	CH$_2$CH$_2$—⬡(OMe)—OMe	6.8	111	CH$_2$NHSO$_2$Me	8.9
			112	CH$_2$CH$_2$NHSO$_2$Me	8.1
GR 113808		9.0			

[a]Reference 70. [b]Antagonism of 5-HT induced relaxation of rat, carbachol contracted esophagus muscularis mucosae.
[c]Partial agonist.

TABLE 1.12. SAR of 2-benzyloxyaryl ketone 5-HT$_4$ receptor antagonists [a]

compd.	X	R	pK$_i$[b]	pK$_b$[c]
113	OMe	butyl-*n*	9.32	8.9
114	OMe	pentyl-*n*	9.09	9.1
115	H	CH$_2$CH$_2$NHSO$_2$Me	8.75	8.1
95	OMe	CH$_2$CH$_2$NHSO$_2$Me	9.13	9.2
116	OMe	CH$_2$CH$_2$NMeSO$_2$Me	9.33	9.1
117	OMe	CH$_2$CH$_2$CH$_2$NHSO$_2$Me	9.47	8.8
GR 113808			10.20	9.0
SB 204070			10.89	10.3

[a]Reference 70. [b]Displacement of [^{3}H]GR 113808 from guinea-pig striatal membranes. [c]Antagonism of 5-HT mediated relaxation of rat, carbachol contracted esophageal muscularis mucosae.

TABLE 1.13. SAR of benzodioxanyl ketone 5-HT$_4$ receptor antagonists [a]

compd.	R	pK$_b$[b]
118	Bu-*n*	9.9
119	CH$_2$CH$_2$NHSO$_2$Me	9.9
120	CH$_2$CH$_2$CH$_2$Ph	9.6
121	CH$_2$CH$_2$CH$_2$C$_6$H$_4$OMe-4	10.6
96 (RS-100235)	CH$_2$CH$_2$CH$_2$C$_6$H$_3$(OMe)$_2$-3,4	11.2
SB 204070		10.3
GR 113808		9.0

[a]Reference 71. [b]Antagonism of 5-HT induced relaxation of rat, carbachol contracted esophagus muscularis mucosa

TABLE 1.14. 5-HT$_4$ receptor antagonist activity of azabicylic indole esters[x]

compd.	R	pIC$_{50}$[b]
124		5.8
125		8.5
126		8.2
127		5.8
128		5.6
129		7.4
130		8.1
131		7.7
132		8.4
133		8.1

[a]Reference 76. [b]Inhibition of 5-HT evoked contractions in the guinea pig distal colon.

TABLE 1.15. 5-HT$_4$ receptor antagonist activity of indole esters and amides

compd.		pIC$_{50}$[a]	pK$_i$[b]	ref.
136	X = O	9.3	8.7	76,77
137	X = NH	6.7	—	77
138	X = O	10.0	—	77
139	X = NH	7.1	—	77
140	X = O	10.0	10.1	76,77
141	X = NH	7.8	—	77
142	X = O	10.6	10.2	76,77
143	X = NH	9.2	9.3	77
144	X = O	9.8	9.7	76,77
145	X = NH	8.1	—	77
146		9.5	10.0	76
SB 204070 (**11**)		10.1	9.9	28

[a]Inhibition of 5-HT evoked contractions in the guinea pig distal colon. [b]Displacement of [^{125}I]SB 207710 from piglet hippocampal membranes.

TABLE 1.16. 5-HT$_4$ receptor antagonist activity of carbazimidamides[a]

compd.	R[b]	R$_1$	pIC$_{50}$[c]
74		H	—[d]
151		H	8.0
152		Me	8.4
153		Me	6.8
154		H	8.8
155		Me	7.1

[a]Reference 85. [b]Indol-3-yl group. [c]Inhibition of 5-HT evoked contractions in the guinea ileum. [d]Agonist, Table 1.9.

TABLE 1.17. 5-HT$_4$ receptor antagonist activity of phenylcarbamates and phenylureas[a]

compd.	X	Y	Z	pK$_b$[b]
156	H	O	NHSO$_2$Me	10.8
157	5-F	O	NHSO$_2$Me	8.9
158	4-OMe	O	NHSO$_2$Me	8.1
159	H	O	N(Me)SO$_2$Me	8.6
160	H	O	OMe	10.0
161	H	NH	NHSO$_2$Me	7.8

[a]Reference 86. [b]Inhibition of 5-HT-induced relaxation of the rat esophagus.

TABLE 1.18. 5-HT$_4$ receptor activity of phenylcarbamates^a

compd.	X	Y	R	R$_1$	pK$_i^b$	pIC$_{50}^c$
162	Cl	NH$_2$	Me	H	7.4	—
163	Cl	NH$_2$	Me	Me (cis)	8.6	6.7
164	H	H	Me	H	8.0	6.5
165	H	H	Et	H	8.6	7.2
166	Cl	H	Me	H	8.4	7.5
167	H	H	Me	Me (cis)	8.6	6.8
168	H	H	Et	Me (cis)	9.0	7.5
169	H	H	Et	Me (trans)	8.9	7.0
GR 113808					9.8	7.9

^{a}Reference 87. ^{b}Displacement of [^{3}H]GR 113808 from rat striatal membranes. ^{c}Inhibition of 5-HT-evoked contractions in the electrically stimulated guinea pig ileum.

References

1. Dumuis A, Bouhelal R, Sebben M et al. A nonclassical 5-hydroxytryptamine receptor positively coupled with adenylate cyclase in the central nervous system. Mol Pharmacol 1988; 34:880-887.
2. Dumuis A, Bouhelal R. Sebben M et al. A 5-HT receptor in the central nervous system positively coupled to adenylate cyclase is antagonized by ICS 205-930. Eur J Pharmacol 1988; 146:187-188.
3. Bockaert J, Fozard JR, Dumuis A et al. The 5-HT$_4$ receptor: a place in the sun. Trends Pharmacol Sci 1992; 13:141-145.
4. Ford APDW, Clarke DE. The 5-HT$_4$ receptor. Med Res Rev 1993; 13:633-662.
5. Eglen RM, Hegde SS. 5-Hydroxytryptamine (5-HT$_4$) receptors: physiology, pharmacology and therapeutic potential. Exp Opin Invest Drugs 1996; 5:373-388.
6. Dumuis A, Sebben M, Boeckaert. The gastrointestinal prokinetic benzamide derivatives are agonists at the non-classical 5-HT receptor (5-HT$_4$) positively coupled to adenylate cyclase in neurons. Naunyn-Schmiedeberg's Arch Pharmacol 1989; 340:403-410.

7. Bockaert J, Sebben M, Dumuis A. Pharmacological characterization of 5-hydroxytryptamine₄ (5-HT₄) receptors positively coupled to adenylate cyclase in adult guinea pig hippocampal membranes: effects of substituted benzamide derivatives. Mol Pharmacol 1990; 37:408-411.

8. King FD. Structure activity relationships of 5-HT₃ receptor antagonists. In: King FD, Jones BJ, Sanger GJ, eds. 5-Hydroxytryptamine-3 Receptor Antagonists. Boca Raton: CRC Press, 1994:1-44.

9. Flynn DL, Zabrowski DL, Becker DP et al. SC-53116: The first selective agonist at the newly identified serotonin 5-HT₄ receptor subtype. J Med Chem 1992; 35:1486-1489.

10. Gaster LM, King FD. Sertonin 5-HT₃ and 5-HT₄ receptor antagonists. Med Res Rev 1997; 2:163-214.

11. Craig DA, Clarke DE. Pharmacological characterization of a neuronal receptor for 5-hydroxytryptamine in guinea pig ileum with properties similar to the 5-hydroxytryptamine₄ receptor. J Pharmacol Exp Ther 1990; 252:1378-1386.

12. Wardle KA, Sanger GJ. The guinea-pig distal colon—a sensitive preparation for the investigation of 5-HT₄ receptor-mediated contractions. Br J Pharmacol 1993; 110:1593- 1599.

13. Baxter GS, Craig DA, Clarke DE. 5-Hydroxytryptamine₄ receptors mediate relaxation of the rat oesophageal tunica muscularis mucosae. Naunyn-Schmiedeberg's Arch Pharmacol 1991; 343:439-446.

14. Grossman CJ, Kilpatrick GJ, Bunce KT. Development of a radioligand binding assay for 5- HT₄ receptors in guinea-pig and rat brain. Br J Pharmacol 1993; 109:618-624.

15. Brown AM, Young TJ, Patch TL et al. [¹²⁵I]-SB 207710, a potent, selective radioligand for 5-HT₄ receptors. Br J Pharmacol 1993; 110:10P.

16. Droppleman DA, Gregory RL, Alphin RS. A simplified method for assesssing drug effects on gastric emptying in rats. J Pharmacol Meth 1980; 4:227-230.

17. Banner SE, Smith MI, Sanger GJ. 5-HT receptors and 5-hydroxy-tryptophan-evoked defacation in mice. Br J Pharmacol 1993; 110:135P.

18. Bingham S, King BF, Rushant B et al. Antagonism by SB 204070 of 5-HT-evoked contractions in the dog stomach: an in-vivo model of 5-HT₄ receptor function. J Pharm Pharmacol 1995; 47:219-222.

19. Banner SE, Smith MI, Bywater D et al. 5-HT₄ receptor antagonism by SB 204070 inhibits 5-hydroxytryptophan-evoked defaecation in mice. Br J Pharmacol 1993; 110:17P.

20. Hegde SS, Moy TM, Perry M et al. Evidence for the involvement of 5-HT₄ receptors in 5- hydroxytryptophan-induced diarrhea in mice. J Pharmacol Exp Ther 1994; 271:741-747.

21. Villalon CM, Den Boer MO, Heligers JP et al. Further characterization, by the use of tryptamine and benzamide derivatives, of the putative 5-HT₄ receptor mediating tachycardia in the pig. Br J Pharmacol 1991; 102:107-112.

22. Eglen RM, Alvarez R, Johnson LG et al. The action of SDZ 205,557 at 5-hydroxytrypamine ($5-HT_3$ and $5-HT_4$) receptors. Br J Pharmacol 1993; 108:376-382.

23. Fontana DJ, Daniels SE, Wong EHF et al. The effects of novel, selective 5-hydroxytryptamine ($5-HT_4$) receptor ligands in rat spatial navigation. Neuropharmacol 1997; 36:689-696.

24. Buchheit KH, Gamse R, Pfannkuche HJ. SDZ 205-557, a selective, surmountable antagonist for $5-HT_4$ receptors in the isolated guinea pig ileum. Naunyn-Schmiedeberg's Arch Pharmacol 1992; 345:387-393.

25. Susemichel AD, Sandusky GE, Cohen ML. 2nd International symposium on serotonin 1992. Houston, Abst. P49.

26. Eglen RM, Bley K, Bonhaus DW et al. RS-23597-190: a potent and selective $5-HT_4$ receptor antagonist. Br J Pharmacol 1993; 110: 119-126.

27. Gale, JD, Grossman CJ, Whitehead JWF et al. GR113808: a novel, selective antagonist with high affinity at the $5-HT_4$ receptor. Br J Pharmacol 1994; 111: 332-338.

28. Gaster LM, Jennings AJ, Joiner GF et al. (1-Butyl-4-piperidinyl)methyl 8-amino-7-chloro-1,4-benzodioxane-5-carboxylate hydrochloride: a highly potent and selective $5-HT_4$ receptor antagonist derived from metoclopramide. J Med Chem 1993; 36:4121-4123.

29. Dumuis A, Sebben M, Monferini E et al. Azabicycloalkyl benzimidazolone derivatives as a novel class of potent agonists at the $5-HT_4$ receptor positively coupled to adenylate cyclase in brain. Naunyn-Schmiedeberg's Arch Pharmacol 1991; 343:245-251.

30. Turconi m, Schiantarelli P, Borsini F et al. Azabicycloalkyl benzimidazolones: interaction with serotonergic $5-HT_3$ and $5-HT_4$ receptors and potential therapeutic implications. Drugs of the Future 1991; 16:1011-1026.

31. Schiavone A, Giraldo E, Giudici L et al. DAU 6285: a novel antagonist at the putative $5-HT_4$ receptor. Life Sci 1992; 51:583-592.

32. Croci T, Langlois M, Mennini T et al. ML 10302, a powerful and selective new $5-HT_4$ receptor agonist. Br J Pharmacol 1995; 114: 382P.

33. Clark RD, Jahangir A, Langston JA et al. Ketones related to the benzoate $5-HT_4$ receptor antagonist RS-23597 are high affinity partial agonists. Bioorg Med Chem Lett 1994; 4:2477-2480.

34. Elz S, Keller A. Preparation and in vitro pharmacology of $5-HT_4$ receptor ligands. Partial agonism and antagonism of metoclopramide analogous benzoic esters. Arch Pharm 1995; 328:585-594.

35. Langois M, Zhang L, Yang D et al. Design of a potent $5-HT_4$ receptor agonist with nanomolar affinity. Bioorg Med Chem Lett 1994; 4:1433-1436.

36. Yang D, Soulier JL, Sicsic et al. New esters of 4-amino-5-chloro-2-methoxlybenzoic acid as potent agonists and antagonists for $5-HT_4$ receptors. J Med Chem 1997; 40:608-621.

37. Hadley MS, King FD, McRitchie B et al. Substituted benzamides with conformationally restricted side chains. 1. Quinolizidine derivatives as selective gastric prokinetic agents. J Med Chem 1985; 28:1843-1847.

38. Hadley MS, King FD, McRitchie B et al. Substituted benzamides with conformationally restricted side chains. 3. Azabicyclo[x.y.o] derivatives as gastric prokinetic agents. Bioorg Med Chem Lett 1992; 2:1147-1152.

39. Hadley MS, King FD, McRitchie B et al. Substituted benzamides with conformationally restricted side chains. 4. Hetero-azabicyclo[x.y.o] derivatives as gastric prokinetic agents. Bioorg Med Chem Lett 1992; 2:1293-1298.

40. King FD, Hadley MS, Joiner KT et al. Substituted benzamides with conformationally restricted side chains. 5. Azabicyclo[x.y.z] derivatives as 5-HT₄ receptor agonists and gastric motility stimulants. J Med Chem 1993; 36:683-689.

41. Baxter GS, Boyland P, Gaster LM et al. Quaternized renzapride as a potent and selective 5-HT₄ receptor agonist. Bioorg Med Chem Lett 1993; 3:633-634.

42. Becker DP, Goldstin B, Gullikson GW et al. Design and synthesis of agonists and antagonists of the serotonin 5-HT₄ receptor subtype. Pharmacochem Libr (Perspectives in Receptor Research) 1996; 24:99-120.

43. Flynn DL, Becker DP, Spangler DP et al. New aza(nor)adamantanes are agonists at the newly identified serotonin 5-HT₄ receptor and antagonists at the 5-HT₃ receptor. Bioorg Med Chem Lett 1992; 2:1613-1618.

44. Gullikson GW, Virina MA, Loeffler RF et al. SC-49518 enhances gastric emptying of solid and liquid meals and stimulates gastrointestinal motility in dogs by a 5-hydroxytryptamine₄ receptor mechanism. J Pharmacol Exp Ther 1993; 264:240-248.

45. Langlois M, Yang D, Bremont B et al. Synthesis and pharmacological activity of a macrocyclic benzamide. Bioorg Med Chem Lett 1995; 5:795-798.

46. Clark RD, Weinhardt KK, Berger J et al. N-(quinuclidin-3-yl)1,8-naphthalimides with 5-HT₃ receptor antagonist and 5-HT₄ receptor agonist properties. Bioorg Med Chem Lett 1993; 3:1375-1378.

47. Eglen RM, Bonhaus DW, Clark RS et al. (R) and (S) RS 56532: mixed 5-HT₃ and 5-HT₄ receptor ligands with opposing enantiomeric selectivity. Neuropharmacol 1994; 33:515-526.

48. Fontana DJ, Wong EHF, Clark R et al. Pro-cognitive effects of RS-66331, a mixed 5-HT₃ receptor antagonist/5-HT₄ receptor agonist. Proceedings of the Third IUPHAR Satellite Meeting on Serotonin. Chicago 1994:57:abst 1.

49. Eglen, RM, Wong EHF, Dumois et al. Central 5-HT₄ receptors. Trends Pharmacol Sci 1995; 16:391-398.

50. Fancelli D, Caccia C, Fornaretto MG et al. Serotonergic 5-HT₃ and 5-HT₄ receptor activities of dihydrobenzofuran carboxylic acid derivatives. Bioorg Med Chem Lett 1996; 6:263-266.

51. Blum E, Buchheit KH, Buescher HH et al. Design and synthesis of novel ligands for the 5-HT₃ and 5-HT₄ receptor. Bioorg Med Chem Lett 1992; 2:461-466.

52. Fernandez AG, Kelly ME, Puig J et al. LAS Z-019, a new centrally acting 5-HT$_4$ agonist. Methods Find Exp Clin Pharmacol 1994; 16, suppl 1: 99.

53. Briejer, MR, Akkermans LMA, Meulemans AL et al. Cisapride and a structural analog, R 76186 are 5-hydroxytryptamine4 (5-HT$_4$) receptor agonists on the guinea pig colon ascendens. Naunyn-Schmiedeberg's Arch Pharmacol 1993; 347:464-470.

54. Kawakita T, Kuroita T, Murozone T. New benzoic acid derivatives are 5-HT$_4$ receptor agonists. Int Pat Appl 1997; WO 97/006452.

55. Yoshida, N. Pharmacological studies on mosapride citrate (AS-4370), a gastroprokinetic agent. (1). Comparison with cisapride and metoclopramide. Yakuri to Chiryo 1993; 21:3013-3028. Chem Abst 120:95340 (1997).

56. Matsuyama S, Sakiyama H, Nei K. Identification of putative 5-hydroxytryptamine$_4$ (5-HT$_4$) receptors in guinea pig stomach: the effect of TKS159, a novel agonist, on gastric motility and acetylcholine release. J Pharmacol Exp Ther 1996; 276:989-995.

57. Eglen RM, Bonhaus DW, Johnson LG et al. Pharmacological characterization of two novel and potent 5-HT$_4$ receptor agonists, RS 67333 and RS 67506, in vitro and in vivo. Br J Pharmacol 1995; 115:1387-1392.

58. Buchheit KH, Gamse R, Giger R et al. The serotonin 5-HT$_4$ receptor. 1. Design of a new class of agonists and receptor map of the agonist recognition site. J Med Chem 1995; 38:2326-2330.

59. Buchheit KH, Gamse R, Giger R et al. The serotonin 5-HT$_4$ receptor. 2. Structure-activity studies of the indole carbazimidamide class of agonists. J Med Chem 1995; 38:2331-2338.

60. Appel S, Kumle A, Hubert M et al. First pharmacokinetic-pharmacodynamic study in humans with a selective 5-hydroxytryptamine$_4$ receptor agonist. J Clin Pharmacol 1997; 37:229-237.

61. Appel, S, Lemarechal MO, Kumle A et al. Simultaneous p.o. and i.v. pharmacokinetic modeling of the novel selective 5-HT$_4$ receptor agonist SDZ HTF 919 in humans. 98th Ann Meeting Am Soc Clin Pharmacol Ther, San Diego 1997; poster Pl-85.

62. Romanelli MN, Ghelardini C, Die S et al. Synthesis and biological activity of a series of aryl tropanyl esters and amides chemically related to 1H-indole-3-carboxylic acid endo 8-methyl-8-azabicyclo[3.2.1]oct-3-yl ester. Arzneim-Forsch/Drug Res 1993; 43:913-918.

63. Ramirez MJ, Garcia-Garayoa E, Monge A. VB20B7, a novel 5-HT-ergic agent with gastrokinetic activity. I. Interaction with 5-HT$_3$ and 5-HT$_4$ receptors. J Pharm Pharmacol 1997; 49:58-65.

64. Gaster LM, Sanger GJ. SB 204070: 5-HT$_4$ receptor antagonists and their potential therapeutic utility. Drugs of the Future 1994; 19:1109-1121.

65. Gaster LM. 8th RSC-SCI medicinal chemistry symposium. Cambridge. 1995.

66. Wardle KA, Ellis ES, Baxter GS et al. The effects of SB 204070, a highly potent and selective 5-HT$_4$ receptor antagonist, on guinea-pig distal colon. Br J Pharmacol 1994; 112:789-794.

67. Hegde SS, Bonhaus DW, Johnson LG et al. RS-39604: a potent, selective and orally active 5-HT$_4$ receptor antagonist. Br J Pharmacol 1995; 115:1087-1095.
68. Yang DC, Goldstin B, Moorman AE et al. SC-53606, a potent and selective antagonist of 5-hydroxytryptamine$_4$ receptors in isolated rat esophagial tunica muscularis mucosae. J Pharmacol Exp Ther 1993; 266:1339-1347.
69. Orjales A, Alonso-Cires L, Labeaga L et al. New substituted benzamides as 5-HT$_4$ receptor antagonists. Eur J Med Chem 1995; 30:651-654.
70. Clark RD, Jahangir A, Langston JA et al. Synthesis and preliminary pharmacological evaluation of 2-benzyloxy substituted aryl ketones as 5-HT$_4$ receptor antagonists. Bioorg Med Chem Lett 1994; 4:2481-2484.
71. Clark RD, Jahangir A, Flippin LA et al. RS-100235: a high affinity 5-HT$_4$ receptor antagonist. Bioorg Med Chem Lett 1995; 5:2119-2122.
72. Gale JD, Grossman CJ, Darton J et al. GR125487: a selective and high affinity 5-HT$_4$ receptor antagonist. Br J Pharmacol 1994; 113:120P.
73. Gale JD, Green A, Darton J et al. GR125487: a 5-HT$_4$ receptor antagonist with a long duration of action in vivo. Br J Pharmacol 1994; 113:119P.
74. King DF, Gaster LM, Mulholland KR. 5-HT$_4$ receptor antagonists. Int Pat Appl 1994; WO 94/27987.
75. Parker SG, Hamburger S, Taylor EM et al. SB203186, a potent 5-HT$_4$ receptor antagonist, in porcine sinoatrial and human and porcine atriuim. Br J Pharmacol 1993; 108:68P.
76. Wyman PA, Gaster LM, King FD et al. Azabicyclo indole esters as potent 5-HT$_4$ receptor antagonists. Bioor Med Chem 1996; 4:255-261.
77. Gaster LM, Joiner GF, King FD et al. *N*-[(1-butyl-4-piperidinyl)methyl]-3,4-dihydro-2*H*- [1,3]oxazino[3,2-a]indole-10-carboxamide hydrochloride: The first potent and selective 5- HT$_4$ receptor antagonist amide with oral activity. J Med Chem 1995; 38:4760-4763.
78. Gaster LM, Wyman PA, Ellis ES et al. 5-HT$_4$ receptor antagonists: oxazolo, oxazino and oxazepino[3,2-a]indole derivatives. Bioorg Med Chem Lett 1994; 4:667-668.
79. Wardle KA, Bingham S, Ellis ES et al. Selective and functional 5-hydroxytryptamine$_4$ receptor antagonism by SB 207266. Br J Pharmacol 1996; 118:665-670.
80. King FD. Modulators of 5-HT function in the treatment of gastrointestinal disorders. 11th Noordwijkerhout-Camerino symposium-Trends in Drug Research, Noordwijkerhout, The Netherlands 1997: lecture 18.
81. Kennet GA, Bright F, Trail B et al. Anxiolytic-like actions of the selective 5-HT$_4$ receptor antagonists SB 204070A and SB 207266A in rats. Neuropharmacol 1997; 36:707-712.
82. Kaumann AJ, King FD, Young RC et al. Indazole as an indole bioisostere: 5-HT$_4$ receptor antagonism. Bioorg Med Chem Lett 1992; 2:419-420.

83. Cohen ML, Bloomquist W, Schaus JM et al. LY353433, a potent, orally effective, long-acting 5-HT$_4$ receptor antagonist: Comparison to cisapride and RS-23597-190. J Pharmacol Exp Ther 1996; 277:97-104.

84. Hagihara K, Hayakawa T, Arai T et al. Antagonistic activities of N-3389, a newly synthesized diazabicyclo derivative, at 5-HT$_3$ and 5-HT$_4$ receptors. Eur J Pharmacol 1994; 271:159-166.

85. Buchheit KH, Klein F, Kloppner E et al. The serotonin 5-HT$_4$ receptor: Part 3: Design and pharmacological evaluation of a new class of antagonists. Bioorg Med Chem Lett 1995; 5:2495-2500.

86. Oxford AW. Substituted phenylcarbamates and phenylureas, their preparation and their use as 5-HT antagonists. Int Pat Appl 1993; WO 93/20071.

87. Soulier JL, Yang D, Bremont B et al. Arylcarbamate derivatives of 1-piperidineethanol as potent ligands for 5-HT$_4$ receptors. J Med Chem 1997; 40:1755-1761.

88. Gaster LM, Joiner GF, Mulholland KR et al. Preparation of N-alkylpiperidinyl-4-methyl carboxylic esters/amides of condensed ring systems as 5-HT$_4$ receptor antagonists. Int Pat Appl 1993; WO 93/20071.

89. Cereda E, Bignotti M, Martino V et al. Preparation of 1-substituted-piperidin-4-ylmethyl esters and amides as 5-HT$_4$ antagonists. Int Pat Appl 1996; WO 96/28424.

90. Sanger GJ, King FD, Baxter GS. Use of heterocyclic derivatives as 5-HT$_4$ receptor antagonists- for treating or preventing gastrointestinal, cardiovascular and central nervous system disorders. Int Pat Appl 1993; WO 93/02677.

91. Jegham S, Galli F, Nedelec A et al. New 5-phenyl-3-(piperidin-4-yl)-1,3,4-oxadiazol-2(3H)-one derivatives are 5-HT4 or 5-HT3 receptor ligands. Int Pat Appl 1997;WO 97/006383.

92. Jegham S, Lochead A, Galli F et al. New 5-phenyl-3-(piperidin-4-yl)-1,3,4-oxadiazol-2(3H)-one derivatives. Int Pat Appl 1997;WO 97/006995.

93. Leung E, Pulido-Rios MT, Bonhaus DW et al. Comparison of 5-HT$_4$ receptors in guinea-pig colon and rat oesophagus: effects of novel agonists and antagonists. Naunyn- Schmiedeberg's Arch Pharmacol 1996; 354:145-156.

94. Bonhaus DW, Berger J, Adham N et al. [^{3}H]RS 57639, a high affinity, selective 5-HT$_4$ receptor partial agonist, specifically labels guinea-pig striatal and rat cloned (5-HT$_{4S}$ and 5 HT$_{4L}$) receptors. Neuropharmacol 1997; 36:671-679.

CHAPTER 2

Molecular Biology of 5-HT$_4$ Receptors

Theresa A. Branchek, Nika Adham, Christophe Gerald

Introduction

Serotonin is a neurotransmitter involved in a plethora of physiological functions exerted through its interaction with numerous receptor subtypes throughout the body. Using pharmacological approaches, four distinct receptor classes were defined: 5-HT$_1$, 5-HT$_2$, 5-HT$_3$ and 5-HT$_4$.[27] However, molecular biological studies have provided both primary amino acid sequence and signal transduction data for a much larger than anticipated array of serotonin receptor subtypes, comprising seven major families: five 5-HT$_1$, three 5-HT$_2$, one 5-HT$_3$, one 5-HT$_4$, two 5-HT$_5$, one 5-HT$_6$ and one 5-HT$_7$ receptor. The majority of these receptors are members of the G-protein-coupled receptor superfamily, which are characterized by a seven transmembrane (TM) spanning arrangement of α helices, proposed to form a barrel-like structure in the membrane.[8] The cloned 5-HT$_1$ receptors are: 5-HT$_{1A}$,[19,31] 5-HT$_{1B}$,[1,14,28,37,52,54] 5-HT$_{1D}$,[11,25,54] 5-HT$_{1E}$[33,39,55] and 5-HT$_{1F}$.[2,6,35] The major signal transduction pathway for each subtype is the inhibition of adenylate cyclase activity. The 5-HT$_2$ receptors include: 5-HT$_{2A}$,[45] 5-HT$_{2B}$[20,32] and 5-HT$_{2C}$.[29] The primary signal transduction pathway of each of these receptors is phosphoinositide hydrolysis. The 5-HT$_3$ receptor is not a G-protein-coupled receptor, it is a ligand-gated ion channel.[36] The 5-HT$_{5A}$ and 5-HT$_{5B}$ receptors are similar in pharmacological profile, but not primary structure, to some 5-HT$_1$ receptors.[18,38] The signal transduction pathways for these subtypes remain elusive, and the 5-HT$_{5B}$ subtype appears to be a pseudogene in the human.[23] Of the seven families, three signal their activation via the stimulation of cAMP production. The 5-HT$_4$ receptor is a member of this cyclase stimulatory serotonin receptor set.[12,22] Both 5-HT$_6$[40] and 5-HT$_7$ receptors[7,34,44,47,48] are also coupled to stimulation of adenylate cyclase. Although one of the early subtypes to be characterized pharmacologically, the 5-HT$_4$ receptor was the last to be cloned.

Cloning of the Rat 5-HT$_4$ Receptor

In order to clone the 5-HT$_4$ receptor, rat brain cDNA was used as a template for PCR amplification with two degenerate oligonucleotide primers derived from the third and fifth transmembrane (TM) domains which are highly conserved in serotonin receptors.[22] This amplification yielded a 270 bp DNA fragment whose deduced peptide sequence contained a "TM IV-like" domain with low homology to previously isolated serotonin receptors. This fragment was used to isolate two full length cDNA clones that were from a rat brain library: 5-HT$_{4S}$ (5.5 kb) and 5-HT$_{4L}$ (4.5 kb). The deduced amino acid sequences of the clones were 96.1% identical, diverging in the second half of the carboxyl terminii at position 360 (Figs. 2.1 and 2.2). The 5-HT$_{4S}$ and 5-HT$_{4L}$ cDNAs encode proteins of 387 and 406 amino acids, respectively, and most likely arise by alternative splicing of pre-mRNA, since their amino acid sequence is identical between positions 1 and 359. Alternative splicing has been described for other 7 TM receptors.[41,50,51] These divergent carboxyl terminii could lead to differences in G-protein-coupling and/or desensitization characteristics.[41,46] The 5-HT$_{4L}$ clone has four protein kinase C phosphorylation sites, whereas the 5-HT$_{4S}$ clone has only three. The additional potential phosphorylation site for protein kinase C is at position 400. Both 5-HT$_{4S}$ and 5-HT$_{4L}$ receptors display a potential N glycosylation site in their amino terminus, and a potential palmitoylation site at the cysteine 329. Many G-protein-coupled receptors have a cysteine in the same position. It has been speculated that this residue may be involved in the functional coupling of α_2 adrenergic and 5-HT$_{1B}$ receptors.[42,43] Cloned 5-HT$_4$ receptors exhibit low levels of TM amino acid identity (less than 50%) with other serotonin receptors, including those that activate adenylyl cyclase.[7,35,40,44,47] Analysis of dendrograms (Fig. 2.3) by similarity analysis revealed a weak clustering of the 5-HT$_4$ receptor with the 5-HT$_2$ subfamily. Comparison of the rat 5-HT$_4$ sequence was as follows: rat 5-HT$_{1A}$ (70% similarity),[5] rat 5-HT$_{1D}$ (67% similarity),[26] rat 5-HT$_{1B}$ (69% similarity),[1] human 5-HT$_{1E}$ (69% similarity),[55] rat 5-HT$_{1F}$ (68% similarity),[34] rat 5-HT$_{2A}$ (73%

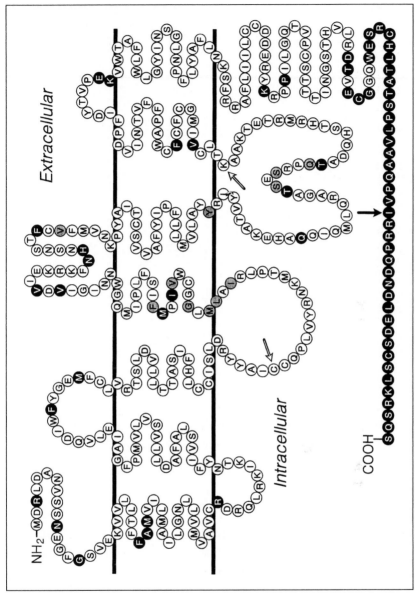

Fig. 2.1. Diagram of the deduced amino acid sequence of the rat 5-HT₄L receptor depicting the regions of transmembrane domains between the dark lines. Residues that differ in either the human, mouse or rat 5-HT₄S sequences are marked in black circles. Regions that differ only in the pig sequence are marked in gray circles. The boundaries of the partial pig sequence are indicated by open arrows. The location of the end of the short form of the rat sequence is marked with a bold arrow.

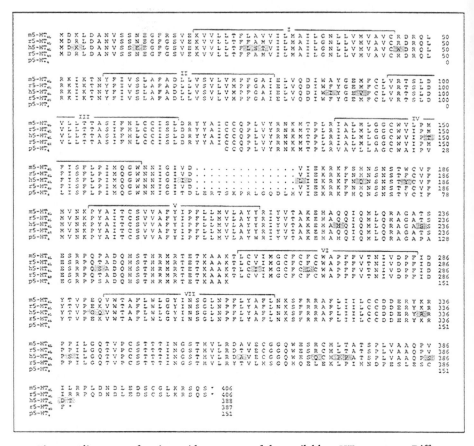

Fig. 2.2. Alignment of amino acid sequences of the available 5-HT₄ receptors. Differences in the human sequences are marked in gray. Genbank accession numbers are: rat 5-HT₄L, U20907; rat 5-HT₄S, U20906; mouse 5-HT₄L, Y09585; pig 5-HT₄, Z48176; human 5-HT₄L from WO 94/14957.

similarity),[30] rat 5-HT₂B (70% similarity),[20] rat 5-HT₂C (77% similarity),[29] rat 5-HT₅A,B (68%, 69% similarity),[18] rat 5-HT₆ (65% similarity)[40] and rat 5-HT₇ (67% similarity).[48]

Cloning of 5-HT₄ Receptors from Other Species

More recently, the mouse 5-HT₄L receptor has been cloned from cultured collicular neurons[13] using PCR amplification. The mouse sequence (Fig. 2.2) predicts a peptide of the identical length as the rat 5-HT₄L, 406 amino acids. A high degree of homology with the rat sequence was noted, giving rise to only 14 differences between the rat and mouse amino acid sequences. The mouse 5-HT₄L had only

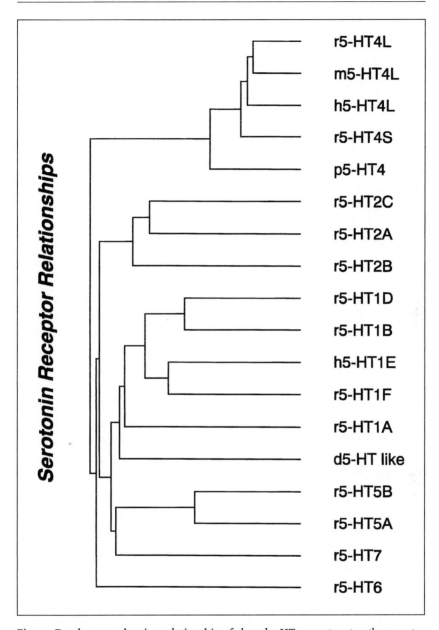

Fig. 2.3. Dendrogram showing relationship of cloned 5-HT₄ receptors to other serotonin receptor subtypes. Known serotonin receptor sequences were compared and clus-
tered according to sequence homology using the Pileup program (Genetics Computer
Group, Inc., Madison, Wisconsin, USA). The lengths of the horizontal lines are inversely
proportional to the sequence homology (similarity, not identity).

two potential phosphorylation sites. These differences may indicate a difference in the potential for desensitization rates between the mouse and rat receptors. This possibility requires further exploration. In the TM domains, there were three differences in the amino acid sequence between mouse and rat: L26F, V28M and L148I. These differences would not predict significant changes in pharmacological profile or G-protein-coupling.

In addition to the mouse 5-HT_{4L} receptor, the human 5-HT_{4L} receptor has also been cloned.[21] Compared to the rat 5-HT_4 sequence, the human 5-HT_{4L} receptor shows 90.7% identity at the nucleotide level and 91.8% identity at the amino acid level (Figs. 2.1 and 2.2). The 5-HT_4 nucleotide sequence contains one nucleotide insertion in position 1159. This insertion creates a frame shift and introduces a stop codon in the reading frame 16 nucleotides downstream. Due to the frame shift, the nucleotide sequence of human 5-HT_{4L} receptor predicts a peptide of 388 amino acids instead of 406 amino acids for the rat 5-HT_{4L} receptor. The protein motifs are highly conserved between the rat and human homologues, except for a casein kinase II potential phosphorylation site in position 288, which is lost in the human receptor and a protein kinase C site formed in position 400 of the rat 5-HT_{4L} sequence, which is lost in the human 5-HT_{4L} receptor. The human homologue carries a potential cAMP/cGMP phosphorylation site in position 338 in its carboxy terminal tail which is absent in the rat homologues. A comparison of the amino acid sequence between the human and the rat clones, beginning from the initiating methionine and ending with the stop codon of the human 5-HT_4 clone, reveals 31 amino acid changes of which 11 are nonconservative, including two in TM1, one in TM2 and one in TM4. Due to the nucleotide insertion and the corresponding frame shift described above, the carboxy terminal tail of the human 5-HT_4 receptor is 16 amino acids shorter than its rat homologue. The human and rat 5-HT_{4L} receptors show six differences in the TM domains: L26F, S27A, T28M (all in TM I), T150M (in TM IV), and I263V, L271F (in TM IV). The carboxyl terminii are 88% identical, differing in 9 out of 72 amino acids between human and rat 5-HT_{4L} receptors. The loop regions are also highly conserved.

A partial sequence of a pig 5-HT_4 receptor has appeared.[49] This sequence begins in the proximal region of the second intracellular loop, includes TM regions IV and V, and ends at the proximal boundary of TM region VI (Figs. 2.1 and 2.2). In the TM region, there are seven unique amino acids in the pig sequence, six of which are in TM IV. In addition, there is a 14 amino acid insertion in the second extracellular loop of the pig sequence which is not observed in mouse, rat or human receptors. It is not clear whether any of these differences lead to interesting pharmacological properties as the full length clone has not been described.

Pharmacological Analysis of the Cloned 5-HT₄ Receptors

The cDNAs encoding the rat 5-HT_{4S} and rat or human 5-HT_{4L} receptors were transiently transfected into COS-7 cells for pharmacological analysis.[10,21,22] For the mouse 5-HT_{4L} receptor, transient transfections were made in LLCPK1 cells.[13] All cloned 5-HT₄ receptors bound [³H]GR113808 with high affinity, and their pharmacological binding profiles obtained from displacement studies were very similar (Table 2.1). The cloned 5-HT₄ receptors are also similar to the native 5-HT₄ receptors as evaluated using functional assays[9,13,17] and radioligand binding assays using [³H]GR113808.[13,24,53] All cloned 5-HT₄ receptors were found to couple to adenylyl cyclase stimulation.

No major differences have been found in receptor pharmacology of the long and short forms of the 5-HT₄ receptor in binding assays using [³H]5-HT or [³H]GR113808 as the radioligand.[4,22] This is not surprising since the amino acid sequences of these two clones are identical, apart from the cytoplasmic carboxy tail, a region that is important for G-protein-coupling.

In functional assays measuring cAMP release, transiently expressed 5-HT_{4S} and 5-HT_{4L} receptors exhibited very similar pharmacological profiles. Some differences were noted; in general, the potency of agonists for stimulating cAMP release was greater for 5-HT_{4S} receptor. For example, the mean EC_{50} values for 5-HT were significantly different (5-HT_{4L}, ≈51 nM; 5-HT_{4S}, ≈25 nM (n = 3), p < 0.05). This difference could be due to higher expression levels (approximately 2-fold) of 5-HT_{4S} receptor as compared to 5-HT_{4L} receptor (B_{max}s of 5-HT_{4L} using [³H]GR113808, ≈3 pmol/mg protein; 5-HT_{4S}, ≈2 pmol/mg protein (n = 3) p < 0.05).[22] Moreover, the maximal stimulation elicited by 5-HT_{4L} receptor was significantly greater than that produced by 5-HT_{4S} (E_{max}s of 5-HT_{4L}, ≈2,500%; 5-HT_{4S}, ≈2,000% basal cAMP release (n = 3) p < 0.05) despite the lower expression levels.[22] These data indicate that 5-HT_{4S} and 5-HT_{4L} receptors may possess different coupling efficiencies to elicit

Table 2.1. Affinities (pKi) for [3H] GR113808 binding

Drug	Rat 5-HT$_{4S}$[10]	Rat 5-HT$_{4L}$[22]	Mouse 5-HT$_{4L}$[13]	Human 5-HT$_{4L}$[21]	Mouse Neurons[13]	Rat Esophagus*
5-HT	6.9	6.84	7.25	7.1	7.1	8.2
5-MeOT	6.3	6.4	6.64	6.76	6.28	8
α-CH$_3$-5-HT	5.6	5.84	NA	5.72	NA	NA
5-CT	<5.5	<6	5	<6	5	NA
cisapride	7.1	6.91	7.49	6.76	7.29	7.3
BRL-24924	6.72	6.61	6.88	6.66	7.01	7.6
zacopride	6.1	6.09		5.97	NA	NA
S-zacopride	NA	NA	6.72	NA	6.65	6.7
tropisetron	6.6	6.26	7.06	6.89	6.8	NA
BIMU 8	7.3	NA	7.39	NA	7.6	7.6
BIMU 1	NA	NA	7.54	NA	7.27	7.9
GR 113808	8.7	NA	9.69	NA	9.5	NA
SDZ 205,557	7.7	NA	8.11	NA	8.2	NA
DAU 6285	7.2	NA	7.26	NA	7.57	NA

*pEC$_{50}$ values[17]

NA = Not Available

functional responses. Whether 5-HT$_{4S}$ and 5-HT$_{4L}$ receptors couple to different isoforms of G$_s$ and/or adenylyl cyclase remains to be investigated.

The rat 5-HT$_{4L}$ sequence has four protein kinase C phosphorylation sites, whereas 5-HT$_{4S}$ sequence only has three.[22] It has been hypothesized that alternative carboxy terminal tails and the additional phosphorylation site could lead to differences in G-protein-coupling and/or desensitization characteristics of these two receptors.[3] The rat 5-HT$_{4S}$ and 5-HT$_{4L}$ receptors were expressed stably in HeLa cells to investigate whether there was a difference in the signaling or desensitization characteristic of these receptors. Two cell lines expressing either receptor were studied with B$_{max}$ values that were not significantly different from each other (5-HT$_{4S}$ B$_{max}$ = 85 fmoles/mg prot; 5-HT$_{4L}$ B$_{max}$ = 45 fmoles/mg prot, n = 3, p = 0.0539) as measured by [^{3}H]GR113808 binding. Each receptor mediated a dose-dependent and pharmacologically specific stimulation of cAMP accumulation, but the two forms of 5-HT$_4$ receptor were distinguished by their desensitization characteristics. Prior treatment of 5-HT$_{4S}$ and 5-HT$_{4L}$ receptor-expressing cells with 5-HT was accompanied by desensitization of the receptor-stimulated cAMP response.

However, the rate of desensitization was faster for the $5\text{-}HT_{4L}$ receptor (half maximal desensitization, 10 min) compared with the $5\text{-}HT_{4S}$ receptor (90 min or longer).[4] The difference in the desensitization properties of $5\text{-}HT_{4S}$ and $5\text{-}HT_{4L}$ receptors may be an important distinguishing feature of these two closely related splice variants and may be a mechanism by which functional diversity is established in native tissues.

The mouse $5\text{-}HT_{4L}$ receptor appeared to couple more robustly to cyclic AMP stimulation than the rat receptor, although this was at least partially attributed to the higher expression level of the mouse receptor in the host cell.[13] Detailed functional analysis of human $5\text{-}HT_{4L}$ receptors has not been reported.

Summary and Future Directions

Although one of the first 5-HT receptors to be detected by second messenger responses,[15,16] $5\text{-}HT_4$ receptors have been the most recent to be cloned. In contrast to many other 5-HT receptor subfamilies, there is no evidence for multiple subtypes of $5\text{-}HT_4$ receptors based on molecular biology approaches. Although the generation of splice variants such as $5\text{-}HT_{4S}$ and $5\text{-}HT_{4L}$ by processes such as alternative splicing may have significant functional consequences in terms of receptor desensitization, exact coupling to intracellular proteins and more subtle aspects of their regulation have not yet been explored in detail. To date, no molecular modeling studies have appeared on the three-dimensional structure of the $5\text{-}HT_4$ receptor, no site-directed mutagenesis explorations to search for residues that may interact with the numerous selective antagonists now available. Additional future studies may also yield the generation of both constitutive and conditional knock-out mice. It will be intriguing to determine whether the phenotype of such animals is similar to animals treated chronically with $5\text{-}HT_4$ antagonists. Finally, with the burgeoning field of genomics, it will be interesting to learn whether naturally occurring mutations in the $5\text{-}HT_4$ gene are associated with any pathophysiological conditions in either humans or animals.

Acknowledgments

The authors thank Ms. Elena Marton for preparing the manuscript, Dr. Thomas Laz for sequence alignments and helpful discussions and Mr. George Moralishvili for preparing the figures.

References

1. Adham N, Romanienko P, Hartig P et al. The rat 5-hydroxytryptamine₁ᵦ receptor is the species homologue of the human 5-hydroxytryptamine₁ᴰᵦ receptor. Mol Pharmacol 1992; 41:1-7.

2. Adham N, Kao HT, Schechter LE et al. Cloning of another human serotonin receptor (5-HT₁ᶠ): A fifth 5-HT₁ receptor subtype coupled to the inhibition of adenylate cyclase. Proc Natl Acad Sci USA 1993; 90:408-412.

3. Adham N, Gerald C, Vaysse PJJ et al. Differential sensitivity of the short and long rat 5-HT₄ receptor subtypes to desensitization. Soc Neurosci Abstr 1995; 21:773.

4. Adham N, Gerald C, Schechter L et al. [³H]5-Hydroxytryptamine labels the agonist high affinity state of the cloned rat 5-HT₄ receptor. Eur J Pharmacol 1996; 304:231-235.

5. Albert PR, Zhon QY, VonTol HHM et al. Cloning, functional expression and mRNA tissue distribution of the rat 5-hydroxytryptamine₁ₐ receptor gene. J Biol Chem 1990; 265:5825-5832.

6. Amlaiky N, Ramboz S, Boschert U et al. Isolation of a mouse "5HT1E-like" serotonin receptor expressed predominantly in hippocampus. J Biol Chem 1992; 267:19761-19764.

7. Bard JA, Zgombick J, Adham N et al. Cloning of a novel human serotonin receptor (5-HT₇) positively linked to adenylate cyclase. J Biol Chem 1993; 268:23422-23426.

8. Baldwin JM. The probable arrangement of the helices in G protein-coupled receptors. EMBO J 1993; 12:1693-1703.

9. Bockaert J, Fozard JR, Dumuis A et al. The 5-HT4 receptor: A place in the sun. Trends Pharmacol Sci 1992; 13:141-145.

10. Bonhaus DW, Berger J, Adham N et al. [³H]RS 57639, a high affinity, selective 5-HT₄ receptor partial agonist, specifically labels guinea-pig striatal and rat cloned (5-HT₄ₛ and 5-HT₄ₗ) receptors. Neuropharmacol 1997; 36:671-679.

11. Branchek TA, Zgombick JM, Macchi MJ et al. Cloning and expression of a human 5-HT₁ᴅ receptor. In: Saxena P, Fozard JR, eds. Serotonin: Molecular Biology, Receptors, and Functional Effects. Basel:Birkhauser 1991:21-32.

12. Branchek TA. 5-HT₄, 5-HT₆, 5-HT₇; molecular pharmacology of adenylate cyclase stimulating receptors. Seminars in The Neurosciences 1995; 7:375-382.

13. Claeysen S, Sebben M, Journot L et al. Cloning, expression and pharmacology of the mouse 5-HT₄ₗ receptor. FEBS Lett 1996; 398:19-25.

14. Demchyshyn L, Sunahara RK, Miller K et al. A human serotonin 1D receptor variant (5-HT₁ᴅᵦ) encoded by an intronless gene on chromosome 6. Proc Natl Acad Sci USA 1992; 89:5522-5526.

15. Dumuis A, Bouhelal R, Sebben M et al. A nonclassical 5-hydroxytryptamine receptor positively coupled with adenylate cyclase in the central nervous system. Mol Pharmacol 1988; 34:880-887.

16. Dumuis A, Sebben M, Bockaert J. BRL 24924: A potent agonist at a non-classical 5-HT receptor positively coupled with adenylate cyclase in colliculi neurons. Eur J Pharmacol 1989; 162:381-384.

17. Eglen RM, Wong EHF, Dumuis A et al. Central 5-HT$_4$ receptors. Trends Pharmacol Sci 1995; 16:391-398.
18. Erlander MG, Lovenberg TW, Baron BM et al. Two members of a distinct subfamily of 5-hydroxytryptamine receptors differentially expressed in rat brain. Proc Natl Acad Sci USA 1993; 90:3452-3456.
19. Fargin A, Raymond JR, Lohse MJ et al. The genomic clone G-21 which resembles a β-adrenergic receptor sequence encodes the 5-HT$_{1A}$ receptor. Nature 1988; 335:358-360.
20. Foguet M, Hoyer D, Pardo LA et al. Cloning and functional characterization of the rat stomach fundus serotonin receptor. EMBO J 1992; 11:3481-3487.
21. Gerald C, Adham N, Vaysse PJJ et al. The 5-HT$_4$ receptor: Molecular cloning and pharmacological characterization of the human 5-HT$_{4L}$ receptor. Soc Neurosci Abstr 1994; 20:1266.
22. Gerald C, Adham, N, Kao HT et al. The 5-HT$_4$ receptor: Molecular cloning and pharmacological characterization of two splice variants. EMBO J 1995; 14(12):2806-2815.
23. Grailhe R, Ramboz U, Hen R. The 5-HT$_5$ receptors: Characterization of the human 5-HT$_{5A}$ receptor; absence of the human 5-HT$_{5B}$ receptor; knockout of the mouse 5-HT$_{5A}$ receptor. Soc Neurosci Abstr 1995; 21:1856.
24. Grossman CJ, Kilpatrick GJ, Bunce KT. Development of a radioligand binding assay for 5-HT$_4$ receptors in guinea pig and rat brain. Br J Pharmacol 1993; 109:618-624.
25. Hamblin MW, Metcalf MA. Primary structure and functional characterization of a human 5-HT$_{1D}$-type serotonin receptor. Mol Pharmacol 1991; 40:143-148.
26. Hamblin MW, McGuffin RW, Metcalf MA et al. Distinct 5-HT$_{1B}$ and 5-HT$_{1D}$ serotonin receptors in rat: structural and pharmacological comparison of the two cloned receptors. Mol and Cell Neurosci 1992; 3:578-587.
27. Humphrey PPA, Hartig P, Hoyer D. A proposed new nomenclature for 5-HT receptors. Trends Pharmacol Sci 1993; 14:233-236.
28. Jin H, Oksenberg D, Ashkenazi A et al. Characterization of the human 5-hydroxytryptamine$_{1B}$ receptor. J Biol Chem 1992; 267:5735-5738.
29. Julius D, MacDermott AB, Axel R et al. Molecular characterization of a functional cDNA encoding the serotonin$_{1C}$ receptor. Science 1988; 241:558-564.
30. Julius D, Huang KN, Livelli TJ et al. The 5HT2 receptor defines a family of structurally distinct but functionally conserved serotonin receptors. Proc Natl Acad Sci 1990; 87:928-932.
31. Kobilka BK, Frielle T, Collins S et al. An intronless gene encoding a potential member of the family of receptors coupled to guanine nucleotide regulatory proteins. Nature 1987; 329:75-79.
32. Kursar JD, Nelson DL, Wainscott DB et al. Molecular cloning, functional expression, and pharmacological characterization of a novel serotonin receptor (5-hydroxytryptamine$_{2F}$) from rat stomach fundus. Mol Pharmacol 1992; 42:549-557.

33. Levy FO, Gudermann T, Birnbaumer M et al. Molecular cloning of a human gene (S31) encoding a novel serotonin receptor mediating inhibition of adenylyl cyclase. FEBS Lett 1992; 296:201-206.

34. Lovenberg TW, Baron BM, De Lecea L et al. A novel adenylyl cyclase-activating serotonin receptor (5-HT₇) implicated in the regulation of mammalian circadian rhythms. Neuron 1993; 11:449-458.

35. Lovenberg TW, Erlander MG, Baron BM et al. Molecular cloning and functional expression of 5-HT₁E-like rat and human 5-hydroxytryptamine receptor genes. Proc Natl Acad Sci USA 1993; 90: 2184-2188.

36. Maricq AV, Peterson AV, Brake AJ et al. Primary structure and functional expression of the 5HT₃ receptor, a serotonin-gated ion channel. Science 1991; 254:432-436.

37. Maroteaux L, Saudou F, Amlaiky N et al. Mouse 5HT₁B serotonin receptor: Cloning, functional expression and localization in motor control centers. Proc Natl Acad Sci USA 1992; 89:3020-3024.

38. Matthes H, Boschert U, Amlaiky N et al. Mouse 5-hydroxytryptamine₅A and 5-hydroxytryptamine₅B receptors define a new family of serotonin receptors: Cloning, functional expression and chromosomal localization. Mol Pharmacol 1993; 43:313-319.

39. McAllister G, Charlesworth A, Snodin C et al. Molecular cloning of a serotonin receptor from human brain (5HT₁E): A fifth 5HT₁-like subtype. Proc Natl Acad Sci USA 1992; 89:5517-5521.

40. Monsma FJ, Shen Y, Ward RP et al. Cloning and expression of a novel serotonin receptor with high affinity for tricyclic psychotropic drugs. Mol Pharmacol 1993; 43:320-327.

41. Namba T, Sugimoto Y, Negishi M et al. Alternative splicing of C-terminal tail of prostaglandin E receptor subtype EP3 determines G-protein specificity. Nature 1993; 365:166-170.

42. Ng GYK, George SR, Zastawny RL et al. Human serotonin₁B receptor expression in Sf9 cells: phosphorylation, palmitoylation and adenylyl cyclase inhibition. Biochemistry 1993; 32:11727-11733.

43. O'Dowd BF, Hnatowich M, Caron MG et al. Palmitoylation of the human beta 2-adrenergic receptor. Mutation of Cys341 in the carboxyl tail leads to an uncoupled nonpalmitoylated form of the receptor. J Biol Chem 1989; 264:7564-7569.

44. Plassat JL, Amlaiky N, Hen R. Molecular cloning of a mammalian serotonin receptor that activates adenylate cyclase. Mol Pharmacol 1993; 44:229-236.

45. Pritchett DB, Bach AWJ, Wozny M et al. Structure and functional expression of cloned rat serotonin 5HT-2 receptor. EMBO J 1988; 7:4135-4140.

46. Reisine T, Kong H, Raynor K et al. Splice variant of the somatostatin receptor 2 subtype, somatostatin receptor 2B, couples to adenylyl cyclase. Mol Pharmacol 1993; 44:1016-1020.

47. Ruat M, Traiffort E, Leurs R et al. Molecular cloning, characterization, and localization of a high-affinity serotonin receptor (5-HT₇)

activating cAMP formation. Proc Natl Acad Sci USA 1993; 90: 8547-8551.

48. Shen Y, Monsma FJ, Metcalf MA et al. Molecular cloning and expression of a 5-hydroxytryptamine₇ serotonin receptor subtype. J Biol Chem 1993; 268:18200-18204.

49. Ullmer C, Schmuck K, Kalkman HO et al. Expression of serotonin receptor mRNAs in blood vessels. FEBS Lett 1995; 370:215-221.

50. Vanetti M, Kouba M, Wang X et al. Cloning and expression of a novel mouse somatostatin receptor. FEBS Lett 1992; 311:290-294.

51. Vanetti M, Vogt G, Hollt V. The two isoforms of the mouse somatostatin receptor (mSSTR2A and mSSTR2B) differ in coupling efficiency to adenylate cyclase and in agonist-induced receptor desensitization. FEBS Lett 1993; 331:260-266.

52. Voigt MM, Laurie DJ, Seeburg PH et al. Molecular cloning and characterization of a rat brain cDNA encoding a 5-hydroxytryptamine₁ᵦ receptor. EMBO J 1991; 10:4017-4023.

53. Waeber C, Sebben M, Grossman C et al. [³H]-GR113808 labels 5-HT₄ receptors in the human and guinea-pig brain. Neuroreport 1993; 4:1239-1242.

54. Weinshank RL, Zgombick JM, Macchi M et al. Human serotonin 1D receptor is encoded by a subfamily of two distinct genes: 5-HT₁ᴅα and 5-HT₁ᴅβ. Proc Natl Acad Sci USA 1992; 89:3630-3634.

55. Zgombick JM, Schechter LE, Macchi M et al. Human gene S31 encodes the pharmacologically defined serotonin 5-hydroxytryptamine₁ᴇ receptor. Mol Pharmacol 1992; 42:180-185.

CHAPTER 3

Localization of 5-HT$_4$ Receptors in Vertebrate Brain and Their Potential Behavioral Roles

Joël Bockaert, Aline Dumuis

Introduction

In 1988, we found in mouse colliculi neurons grown in primary cultures, a 5-HT receptor which stimulated cAMP production and had a pharmacology different from that of the receptors described at that time, i.e., the 5-HT$_1$, 5-HT$_2$, and 5-HT$_3$ receptors. In particular, highly potent and specific 5-HT$_{1,2,3}$ antagonists were unable to block this receptor. We therefore proposed to name it, without any permission, the 5-HT$_4$ receptor.[1] Fortunately, based on their pharmacology, transduction mechanisms, physiology and finally, cloning, it became rapidly evident that this proposition was correct. The 5-HT$_4$ receptor is the only member of its class because it shows little amino acid sequence relationship to all other serotonin receptors, including those positively coupled to the adenylyl cyclase (AC) (the 5-HT$_6$ and 5-HT$_7$ receptors).[2-6]

For many years the existence of brain 5-HT$_4$ receptors could only be demonstrated using functional (increased in neuronal excitability)[7] and second messenger experiments (cAMP production).[1] Using these tests, 5-HT$_4$ receptors were found in prenatal rodent colliculi and adult rodent hippocampus.[1,7,8] However, the latter localization, plus the fact that 5-HT$_4$ receptors were stimulating cAMP production, a second messenger implicated in learning and memory in invertebrates as well as in vertebrates,[9] led us to suggest that these receptors may play a role in cognition.[4] The introduction

5-HT$_4$ Receptors in the Brain and Periphery, edited by Richard M. Eglen.
© 1998 Springer-Verlag and R.G. Landes Company.

of selective radioligands to label 5-HT$_4$ receptors in the brain confirmed their colliculi and hippocampal localizations, but also indicated that they have a wider distribution (for a review, see Eglen et al).[5] An idea of the neuronal compartment in which they are expressed (somato-dendritic versus axonal terminals) has recently been made possible by comparing the distribution of 5-HT$_4$ receptor mRNA with the distribution of the corresponding 5-HT$_4$ protein, whereas lesion studies have given some insight into their cellular localization.

Regional Distribution of 5-HT$_4$ Receptors in Vertebrate Brain

Nature of the Radioligand Used

Two main radioligands have recently been developed to study the regional distribution of 5-HT$_4$ receptors in the brain of different species: an indole carboxylate, the [³H]-GR113808[10] and a benzoate dioxane, the [¹²⁵I]-SB207710.[11] These two compounds are antagonists. Since the binding of agonists are more affected than the binding of antagonists by the interactions of receptors with G-proteins or by receptor-receptor interactions (like dimerization), it will be of interest to compare regional distribution of 5-HT$_4$ receptors using agonist radioligands. In this context, only one study used [³H]-BIMU-1,[12] an agonist of central 5-HT$_4$ receptors.[13] With few exceptions, the distribution and density of [³H]-GR113808 binding sites was generally comparable to that of [³H]-BIMU-1 binding sites. Some areas, (periventricular thalamus, habenula, lateral septum) showed a 2- to 3-fold higher density of sites labeled with [³H]-BIMU-1 than with [³H]-GR113808, whereas in hippocampus (CA3, CA1, CA3 subfields) it was the contrary. The reasons for such differences are unknown, but may be related to a difference in the transduction proteins associated with the different areas. One has to recall that a positive coupling of 5-HT$_4$ receptors with adenylyl cyclase has so far only been described in colliculi neurons and in hippocampus[1,14] and that the nonhydrolyzable analog of GTP, GppNHp, had a different effect on the agonist affinity for receptors expressed in hippocampus and striatum.[10] Obviously, additional comparison between agonist and antagonist binding in brain is required.

Two other agonist radioligands have been shown to label 5-HT$_4$ receptors, [^{3}H]-5-HT[15] and [^{3}H]-RS 57639.[16] However, they have essentially been used on striatal membranes from guinea pig or on transfected cell lines, but not for brain 5-HT$_4$ receptor localization.

General Distribution

Binding of [^{3}H]-GR113808 and [^{3}H]-SB207710 have been performed on isolated membranes and on brain slices of rat[10,12,17-19] mouse,[10,12,17] guinea-pig,[10,12,17,20,21] pig,[22] calf,[20] monkey,[12] and human.[20,21,23] With few exceptions, there is a heterogeneous and comparable distribution in adults of different species, discussed below. The distribution of [^{3}H]-GR113808 in rat brain is the best documented and will be taken as a guideline (Table 3.1; Fig. 3.1).

Olfactory System

The olfactory bulb, the olfactory nucleus and the primary olfactory cortex contained a very low density of sites (< 10 fmol/mg). In contrast, olfactory tubercle and the Islands of Calleja contained the highest density (300-400 fmol/mg protein).

Basal Ganglia

The labeling is high in the whole striatum (caudate-putamen), but a mediolateral and dorsolateral gradient has been found.[17-19] In contrast to Vilaro et al,[18] Compan et al[19] did not find a rostro-caudal gradient in this structure. The fundus striati and nucleus accumbens are characterized by a high density of 5-HT$_4$ receptors, whereas the globus pallidus expresses an intermediate level in adult rat brain.[17] In adult guinea-pig, the globus pallidus expresses as many receptors as the nucleus accumbens.[12,17] In rat nucleus accumbens, the density of [^{3}H]-GR113808 is 3-fold higher in the shell than in the core.[19]

Among the few human brain areas which have been studied, the caudate and the putamen were the areas expressing the highest density of 5-HT$_4$ receptors.[20,21,23]

Amygdala, Septo-Hippocampal System

The septal region, the amygdaloid nuclei and the subfields of hippocampus showed a homogeneous intermediary labeling.[17,19] No differences in the hippocampal subfields were noted in rat, whereas in guinea-pig, the CA2 regions expressed more receptors than the other areas.[17] The human hippocampus expressed an intermediary level of 5-HT$_4$ receptors.[20,21,23]

Table 3.1. *Densities of specific binding sites for [³H]-GR113808 (0.1 nM) in the adult rat*

Regions	Abbreviations	Rat	Regions	Abbreviations	Rat
Olfactory system			Cortical areas		
Anterior olfactory nucleus	AON	28 ± 12	Anterior cingulate cortex	ACg	70 ± 13
Olfactory tubercle	Tu	301 ± 28	Frontoparietal cortex	FrPa	42 ± 14
Primary olfactory cortex	PO	21 ± 11	Posterior cingulate cortex	DCg	89 ± 15
Islands of Calleja	ICj	412 ± 49	Retrosplenial cortex	RSpl	59 ± 13
			Striate cortex (Area 17)	Str17	22 ± 9
Basal ganglia			Striate cortex (Area 18)	Str18	49 ± 15
Accumbens nucleus	Acb	177 ± 17	Temporal cortex (Auditory area)	TeAud	12 ± 6
Globus pallidus	GP	50 ± 17	Entorhinal cortex	Ent	83 ± 19
Ventral pallidum	VP	135 ± 38			
Caudate-putamen	CPu	110 ± 15	Midbrain		
Fundus striati	FStr	225 ± 12	Superficial grey layer of the superior colliculus	SuG	118 ± 14
			Inferior colliculus	IC	21 ± 9
Amygdalia	Amy		Central grey	CG	81 ± 11
Medial amygdaloid nucleus	Me	80 ± 12	Dorsal Raphé nucleus	DRN	127 ± 20
			Linear Raphé nucleus	LRN	50 ± 9
Septal region			Medial geniculate nucleus (dorsal)	MGD	61 ± 22
Lateral septal nucleus	LS	92 ± 25	Substantia nigra (reticular part)	SNr	40 ± 17
Bed nucleus of the stria terminalis	BST	122 ± 21	Substantia nigra (lateral part)	SNl	143 ± 19
			Interpeduncular nucleus	IP	271 ± 38

Region		R
Hippocampus		
Dentate gyrus	DG	114 ± 20
CA1 field (pyramidal layer)	CA1	101 ± 22
CA2 field (pyramidal layer)	CA2	128 ± 17
CA3 field (pyramidal layer)	CA3	121 ± 18
Dorsal subiculum	S	132 ± 17
Hypothalamus		
Ventromedial nucleus	VMN	73 ± 13
Dorsal hypothalamic	DH	134 ± 16
Medial preoptic area	MPO	102 ± 15
Medial mammilliary nucleus	MM	65 ± 5
Thalamus		
Medial habenula	MHb	153 ± 21
Laterodorsal nucleus	LD	49 ± 12
Centromedial nucleus	CM	89 ± 9
Reunions nucleus	Re	99 ± 8
Paraventricular nucleux	PV	55 ± 14
Red nucleus	R	7 ± 7
Pons		
Cuneiform nucleus	Cnf	41 ± 15
Pontine nuclei	Pn	8 ± 4
Nucleus of the solitary tract	Sol	100 ± 16
Dorsal tegmental nucleus	DTg	95 ± 25
Dorsal parabrachial nucleus	DPB	38 ± 15
Medulla oblongata		
Hypoglossal nucleus		125 ± 13
Nucleus of the spinal tract of the trigeminal nerve (Caudal part)	Sp5C	39 ± 9
Medial vestibular nucleus	MVe	44 ± 8
Dorsal nucleus of the vagus nerve	MVe	60 ± 12
Cerebellum	Cereb	
Granular layer		35 ± 17
Spinal cord		
Layers 1 and 2 of Rexed		63 ± 13
Ventral horn		30 ± 14
Pineal gland	Pi	121 ± 21

Data taken from Waeber et al.[17]

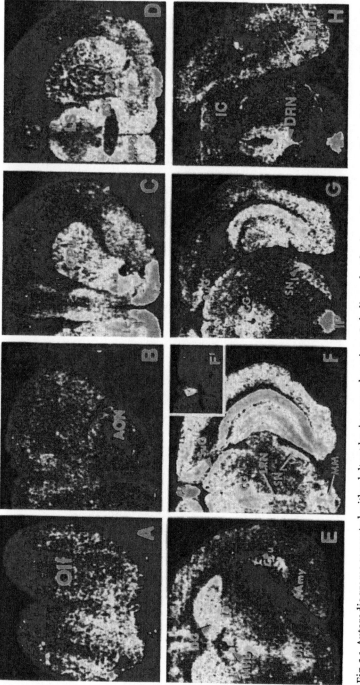

Fig. 3.1. Autoradiograms generated with adult rat brain coronal sections. Labeling of rat brain sections was performed as described by Waeber et al[17] with [³H]-GR113808. Total binding is shown except in F' where 10 mM 5-HT was added to obtain an image of nonspecific binding. The latter is homogeneous in all brain regions except in the pineal gland which exhibits a high level of nonspecific binding. The abbreviations are as in Table 3.1. Photomicrographs taken from Waeber et al.[17]

Hypothalamus and Thalamus

In the hypothalamus, the dorsal hypothalamic nuclei contained the highest density, whereas in rat thalamus only the habenula and in particular, the medial habenula showed a relatively high density.[12,17,18]

Midbrain

In midbrain, the labeling was heterogeneous, ranging from undetectable levels (red nucleus, inferior colliculus) to very high levels (interpeduncular nucleus) via moderate levels (superior colliculus, the periaqueductal central gray and the lateral part of the substantia nigra).[12,17,19] Human substantia nigra expressed a relatively high level of 5-HT₄ receptors.[21] In guinea-pig, as in human, it is the pars reticulata rather than the part lateralis (in rat and mouse) which contained the highest density of this structure.[17,19,21] Similarly, the rat and mouse interpeduncular nucleus is much more labeled than the corresponding structure in guinea-pig.[17] Some raphé nuclei contained intermediate (linear raphé) to high (dorsal raphé) densities.[17]

Cortex and Cerebellum

In the cortex, the concentration of binding sites covered an intermediate range, with a predominant labeling of the cingulate as compared to the fronto-parietal regions with little differences between deeper and upper layers.[17,19] In the cerebellum, the binding was very low.

Hindbrain and Spinal Cord

In the pons, only the dorsal tegmental nucleus, and some cranial nerves (hypoglossal, trigeminal, vagus) expressed an intermediary level of binding sites. In the cervical segment of the spinal cord, the dorsal horn show a clearly higher binding than the ventral horn.[17]

Cellular Localization of Brain 5-HT₄ Receptors

Neuronal Localization

Several indirect observations indicated that 5-HT₄ receptors are essentially located on neurons. Indeed, their distribution is highly heterogeneous in the brain and there is no report which indicates their presence in primary glial cell cultures or in glioma cells.

Locations on GABA and Cholinergic Striatal Neurons

Two studies have used injection of kainic acid in the dorsal part of the striatum, a technique known to destroy neurons having their somato-dendritic compartment within this structure sparing axons terminals and glial cells.[19,24] Both studies indicated that GABA and acetylcholine neurons were largely eliminated by kainic acid, a figure which paralleled the reduction of 5-HT$_4$ receptors in the lesioned area (Fig. 3.2A). These studies also indicated that striatal kainic acid lesions were accompanied by a reduction in 5-HT$_4$ receptor density in the lateral part of the substantia nigra, a well known projection of striatal GABA neurons containing dynorphin and belonging to the cortico-striato-tectal pathway involved in control of saccadic eye movements (Fig. 3.2A). Compan et al also reported that striatal lesions produce an important decrease of the 5-HT$_4$ receptor density in the globus pallidus, which also receives projections from GABA striatal neurons (Fig. 3.2A).[19] Patel et al did not find such a decrease in globus pallidus.[24] However, for an unknown reason they did not find a decrease in the GAD-like immunoreactivity in globus pallidus. In conclusion, 5-HT$_4$ receptors are localized on the somato-dendritic compartment of striatal GABA neurons projecting to lateral substantia nigra and to the globus pallidus. In addition, due to the large decrease (>75%) in 5-HT$_4$ receptor density found within the lesioned area, it is likely that these receptors are also located on intrinsic acetylcholine neurons. These results are in perfect accordance with those of Reynolds et al[23] who found a 50% reduction in 5-HT$_4$ receptor density in postmortem putamen of patients suffering from Huntington's disease in which a profound loss in neurons of basal ganglia is observed.

Location on Glutamatergic Neurons

5-HT$_4$ receptors are located on hippocampal glutamatergic CA1 neurons[7,25] in which they reduced a Ca^{2+}-activated potassium current in these cells.[25] In contrast, they are not located on the cortico-striatal glutamatergic pathway.[19]

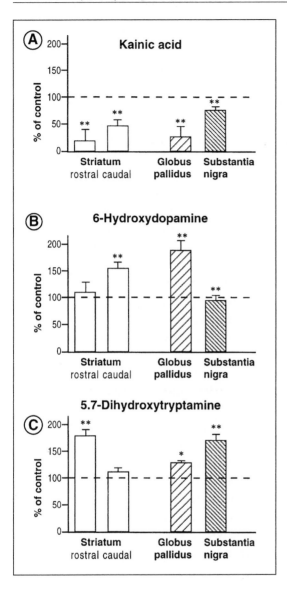

Fig. 3.2. Effects of the neurotoxic lesions with kainic acid, 6-hydroxydopamine, and 5,7-dihydroxytryptamine on 5-HT$_4$ binding in the basal ganglia. Changes in [^{3}H]-GR113808 binding sites in lesioned rat brain areas are expressed as percentages of control values and represent the mean ± SEM of values obtained from five rats per group (six sections per animal). (A) Following kainic acid lesion; (B) After 6-hydroxydopamine lesion of the dopaminergic nigrostriatal pathway; (C) Following the 5,7-dihydroxytryptamine lesion. Differences from controls are statistically significant with $p < 0.05$(*) or $p < 0.01$(**). Data are taken from Compan et al.[19]

Absence of Location on Dopaminergic Neurons

Lesions of nigrostriatal dopaminergic neurons with 6-OH dopamine injected in rat substantia nigra (Fig. 3.2B) and data obtained on substantia nigra of patients suffering from Parkinson's disease (a disease characterized by a specific degeneration of dopamine neurons) clearly indicated that 5-HT$_4$ receptors are not located on dopaminergic neurons.[19,23,24]

Speculations on the Location of 5-HT₄ Receptors on Acetylcholine Neurons

Several reports are in favor of a localization of 5-HT_4 receptors on cholinergic neurons. Among them we can quote: (1) the increase in acetylcholine release from the rat cerebral cortex;[26] (2) the dramatic reduction of 5-HT_4 receptor density in postmortem hippocampus and cerebral cortex of patients suffering from Alzheimer's disease;[23] (3) the well known ability of 5-HT_4 receptors to stimulate acetylcholine release from guinea-pig myenteric plexus; (4) the parallel appearance during the development of 5-HT_4 receptors and choline acetyltransferase in the interpeduncular nucleus, an area receiving a dense cholinergic innervation from medial habenula (in interpeduncular nucleus, the location could be on cholinergic axons terminals of neurons having their cell bodies in the habenula since no 5-HT_4 receptor mRNA was present in this nucleus);[18,27] (5) the reversion by zacopride and renzapride of the decrease in electroencephalographic energy induced by scopolamine.[28] However, the direct demonstration of such a presence is still lacking in brain.

Comparison Between the Regional Distributions of 5-HT₄ Receptors and 5-HT₄ mRNA Further Indicates Somato-Dendritic as well as Axon Terminal Locations

Lesions studies have indicated that 5-HT_4 receptors are expressed in both the somato-dendritic and the axons terminal of the same neurons, for example the striatal GABA neurons.[19,24] Comparison between the distributions of 5-HT_4 receptors and 5-HT_4 mRNA have confirmed this dual compartmentalization.[18,27] Indeed, although the general brain distributions of 5-HT_4 protein and 5-HT_4 mRNA were similar, there was no mRNA in the substantia nigra, globus pallidus, and interpeduncular nucleus, three areas in which an axonal location is likely.

Are 5-HT₄ₛ and 5-HT₄ₗ Receptors Differently Localized in Brain?

Gerald et al[29] cloned two forms of 5-HT_4 receptors differing from the length of their C-termini. They differ downstream of the position 360, the 5-HT_{4S} and 5-HT_{4L} encoding 387 and 406 amino acid protein, respectively. Using RT-PCR analysis, they reported that the mRNA encoding the short form was only expressed in striatum, whereas the mRNA encoding the long form was expressed in all other brain areas including striatum. Using the same technique, we could

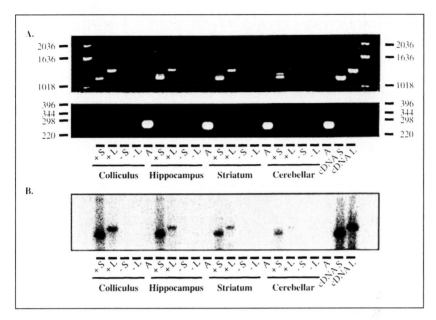

Fig. 3.3. Central distribution of 5-HT$_{4S}$ and 5-HT$_{4L}$ transcripts. RT-PCR analyzes were performed with 100 ng of poly (A+) RNA from various structures of newborn mouse brain. (A) The PCR amplified products were resolved by 1% agarose/ethidium bromide gel electrophoresis, photographed and analyzed by Southern blotting. Upper panel, 5-HT$_{4S}$ and 5-HT$_{4L}$ transcripts; lower panel, new-actin transcripts. (B) Southern blot of new born mouse brain PCR results exposed 90 min to X-ray films. Specific primers based on the known rat cDNA sequences 5-HT$_{4S}$, 5-HT$_{4L}$ or actin were used on poly (A$^+$) RNA treated with (+S, +L, A) or without (-S, -L) reverse transcriptase. A positive control was done by PCR reaction performed with rat cDNA clones encoding 5-HT$_{4S}$ and 5-HT$_{4L}$ receptors. The marker fragments indicated on the right and the left of the photograph show that the size of the PCR products (about 1300 bp) correspond to the total cDNA encoding 5-HT$_{4S}$, and 5-HT$_{4L}$ receptors. Data are taken from Claeysen et al.[30]

not confirm this observation. In contrast, we found that both 5-HT$_{4S}$ and 5-HT$_{4L}$ mRNA were expressed in all brain areas, in rat and in mouse brain, whatever the postnatal development[30] (Fig. 3.3). This finding was confirmed using an in situ hybridization approach and specific probes for 5-HT$_{4S}$ and 5-HT$_{4L}$.[18] In conclusion, the general distributions of 5-HT$_{4S}$ and 5-HT$_{4L}$ mRNA are not different. This does not mean that they are expressed by the same cells within the various brain areas or that a difference between their somato-dendritic versus axonal distribution does not exist.

Ontogenic Development of 5-HT₄ Receptors in Rodent Brain

We have studied the regional distribution of 5-HT_4 receptors in rat and mouse brain at both prenatal and postnatal stages.[17] In embryos, the density of receptors was very low except in the ventral part of the pons. It is noteworthy that a very low level of receptors was observed in superior colliculi before birth, whereas significant densities (> 60 fmol/mg protein) of sites can be detected in mouse neuronal cultures prepared from the same region (at prenatal day 16) and kept for 6 days in vitro.[31] At postnatal day 1, the concentration of receptors increased in most regions to reach a maximal level at postnatal day 12-15. The only exception was in the pons in which the level was constant during the postnatal life and then fell down after 10 days of being negligible in adult. Two other areas indicated a decrease in receptor expression after 15 days of postnatal life, the globus pallidus and the substantia nigra. In the globus pallidus, this peak is synchronous with the transient glutamatergic innervation in this structure. The high density of 5-HT_4 receptors in the brainstem prenatally and their disappearance starting the second week postnatally may indicate a developmental function. A similar decrease of binding sites in the brainstem has been described for substance P, nicotine and GABA-A receptors (see Waeber et al).[17]

Modulation of Central 5-HT₄ Receptors

Upregulation of 5-HT₄ Receptors in Rat Brain, Following Serotoninergic but also Dopaminergic Denervation

In colliculi neurons, as well as in rat esophagus, we described that 5-HT_4 receptors desensitized rapidly.[32,33] During a first period of stimulation with an agonist (which lasted around 20 min), the response decreased at least by 60-70% without any change in receptor density. We have shown that this rapid desensitization was homologous, depending only on receptor occupation without any implication of the intracellular messenger, the cAMP. In colliculi neurons, when the stimulation of the agonist persisted for more than 20 min, a downregulation of the receptors occurred.[31] This indicates that in vivo, the level of expression of 5-HT_4 receptors, may be related to the tonic activation of the receptor. Indeed, lesions of 5-HT neurons with 5,7-dihydroxytryptamine induced an upregulation of 5-HT_4 receptor density in the rostral but not the caudal part of the caudate putamen, in the nucleus accumbens, and in the hippocam-

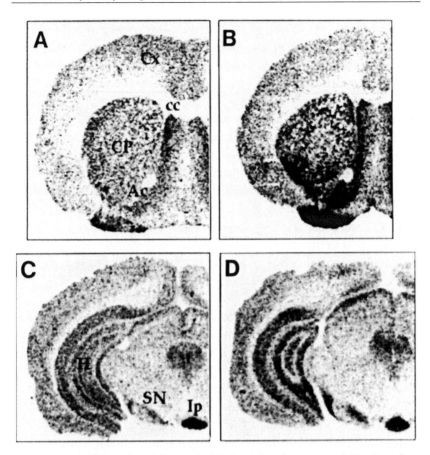

Fig. 3.4. Autoradiographs of the 5-HT$_4$ binding sites from control (A, C) and 5,7 dihydroxytryptamine (B, D) lesioned rat. (A,B) The [³H]-GR 113808 binding is highly increased, following 5,7-dihydroxytryptamine lesion in both the caudate-putamen and the nucleus accumbens. (C,D) Following 5,7-dihydroxytryptamine lesion, the hippocampal and nigral [³H]-GR113808 labeling are increased. The labeling in interpeduncular nucleus (Ip), particularly intense in the control, remains unchanged in the lesioned rats. Data taken from Compan et al.[19]

pus[19] (Figs. 3.2C and 3.4). The reason for the absence of upregulation in the caudal part of the caudate putamen is not clear because these areas received the highest 5-HT innervation. More unexpected was the fact that dopaminergic denervation also induces an upregulation of 5-HT$_4$ receptors in the caudal, but not the rostral part of the caudate-putamen, in the globus pallidus, but not in the substantia nigra (Fig. 3.2).[19] In line with this result, it has recently been published that striatal 5-HT$_{1B}$ and 5-HT$_2$ receptors can be upregulated after dopaminergic denervation.[34] This lesion was followed by considerable sprouting of 5-HT fibers in the caudate-putamen, suggesting that upregulation of 5-HT receptors can be associated with both a 5-HT denervation and hyperinnervation. However, it has also been reported that dopaminergic lesions with 6-OH DA similar to those done in our studies depleted 5-HT stores.[35] 5-HT$_4$ receptors facilitate dopaminergic release in the caudate-putamen.[36-39] Since the 5-HT$_4$ receptors are not on the dopaminergic neurons, the increase in dopamine release mediated by 5-HT$_4$ receptors may be secondary to the activation of striatal neurons. Taken together, these data reinforce the view that complex reciprocal 5-HT / dopamine interactions are involved in the basal ganglia function. The fact that upregulation of 5-HT$_4$ receptors was restricted to the rostral part after 5-HT denervation and to the caudal part after dopaminergic innervation, may reflect differential influences exerted by the two monoaminergic systems in striatal regions.

Subsensitivity of 5-HT$_4$ Receptors Following Repeated Treatment with Antidepressants

In the CA1 cell layer of hippocampus, 5-HT$_4$ agonists, increased the population spikes induced by the stimulation of the Schaffer collateral.[40] This is certainly due to the reduction of the calcium-activated afterhyperpolarization associated with a slow membrane depolarization.[25] It has recently been reported that repeated (14 days, twice daily) administration of antidepressants such as imipramine, citalopram, fluvoxamine, and paroxetine (10 mg/kg) attenuated the effect of zacopride, a 5-HT$_4$ agonist on population spikes.[40]

Behavioral Effects of 5-HT$_4$ Receptors

Cognition

Using patch-clamp recording, we have shown in mouse colliculi neurons that 5-HT$_4$ receptors inhibited a delayed rectified voltage-dependent K$^+$ current.[41,42] This effect was mediated via cAMP production and blocked by PKA inhibitors. In current-clamp experiments, 5-HT$_4$ receptors and cAMP have been shown to slow the falling phase of the action potential, to inhibit the after hyperpolarization (AHP) which follows the action potentials and spike accommodation during a train of action potentials. All these effects can be explained by inhibition of K$^+$ channels including Ca^{2+}-activated K$^+$ channels.[41] In CA1 neurons of hippocampus, activation of 5-HT$_4$ receptors increases membrane excitability by reducing the Ca^{2+}-activated K^+ current responsible for the slow AHP obtained in these cells.[25] In colliculi neurons, a remarkable observation was that after a short application of 5-HT or 8-Br-cAMP (a few seconds) the blockade of K^+ channels persisted for 2 hours and could be prolonged up to 4 hours in the presence of okadaic acid, an inhibitor of phosphatases 1 and 2A.[41] We have shown that a prolonged inhibition of phosphatases were responsible for the long lasting inhibition of K^+ channels following cAMP accumulation in these neurones. Taken together, these results mirror those seen in *Aplysia californica* in which 5-HT-induced cAMP generation inhibit K^+ currents, resulting in a persistent facilitation of evoked transmitter release. These events provide a mechanistic basis for the sensitization of the gill withdrawal reflex in *Aplysia*, an elementary model of learning. For these reasons, but also because 5-HT₄ receptors are able to stimulate acetylcholine release in rat frontal cortex,[26] we proposed several years ago that 5-HT₄ receptors might play a role in cognition. Several recent studies may support this proposal.[43,44]

We studied the effect of BIMU-1, a 5-HT₄ agonist which is also a 5-HT₃ antagonist on social olfactory recognition in rats, a behavior test which has previously been shown to access short term memory and be sensitive to cholinergic drugs.[43] This test is based on the investigation of an unfamiliar juvenile by an adult rat during two distinct 5-minute presentations (P1 and P2). At a 30-minute delay between each presentation, adults recognize the juvenile as indicated by a reduction of the exploration period (P2 compared to P1) of the juvenile by the adult (P2/P1<1) (Fig. 3.5). In contrast, after a 2-hour delay, all the adults have lost their memory of juvenile (P2/P1 = 1) (Fig. 3.5). The recognition of the juvenile is individual-specific, a different juvenile is recognized as a new individual after 30 minutes (P2/P1 = 1). BIMU-1 administered after the first presentation enhanced short term memory (i.e., recognition of the juvenile after a 2-hour delay) (Fig. 3.5). A 5-HT₃ antagonist (ondansetron) had no effect, whereas a highly specific 5-HT₄ antagonist (GR125487; 10 mg/kg) was highly effective in blocking the BIMU-1 effect. This indicates that the positive effect of BIMU-1 on short term memory was due to its 5-HT₄ agonist properties rather than to its 5-HT₃ antagonist properties.[43]

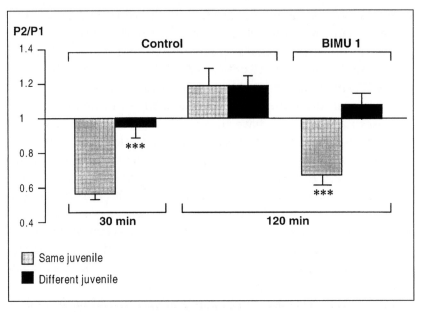

Fig. 3.5. Role of 5-HT$_4$ receptors on social memory. Effect of BIMU-1, a 5-HT$_4$ agonist. Social recognition of juvenile conspecifics by adult rat at a delay of 30 (P2$_{30}$) and 120 min (P2$_{120}$) after P1 ((P1) = first exploration period and (P2) = second exploration period). Investigation times were followed after i.p. administration of BIMU-1 10 mg/kg at a delay of 120 min after the first presentation. y axis: P2/P1, ratio of investigation duration, comparison between the same and different juveniles at P2$_{30}$ and P2$_{120}$ by adult rats after i.p. administration of PBS (control) and at P2$_{120}$ by adult rats after i.p. administration of BIMU-1 10 mg/kg. Data taken from Letty et al.[43]

We have also tested the effect of BIMU-1 on an olfactory associative task.[44] In this task, rats, deprived of water for 48 hours, were trained to make two odor-reward associations. Each odor had to be associated with a specific reward, one arbitrarily designed as positive (S+) and the other as negative (S-). Rats had to approach the odor and water ports to interrupt the light beam only when the positive odor had been delivered. Response to odor designated as negative resulted in presentation of nonadversive light, and water was not delivered. Individual trials were presented in a quasi-random fashion. A daily session was made up 60 trials and five sessions were made with one session per day. Four groups (seven animals each), a saline group, two BIMU-1 (10 mg/kg) groups (in the presence or absence of the antagonist GR125487; 10 mg/kg) and a GR125487 group were used. Drug injections were done before the third session. Whatever the group, their performances were improved across sessions reaching a score superior to 80% on session 5. However, the BIMU-1 group had previously reached this performance, mainly on session 4 (Fig. 3.6A).[44] When the latency to give a response was analyzed, a significant latency difference between S+ and S- was only observed from session 4 for the control group whereas this difference appeared just after the injection on session 3 for the BIMU-1 group (Fig. 3.6B). The performances of the GR125487 and the BIMU-1 plus GR125487 groups were not significantly different from the performance of the control group. A reversal test was performed one month later. Thus, the odor that had served previously as an S+ stimulus in the training session was now an S- stimulus and vice versa. The difficulty of the BIMU-1 group to rapidly reverse their behavioral responses to previously learned associations, one month later (Fig. 3.6C), indicated that the BIMU-1 effect was not transient, but correlated to long-term memory.[44] Again, the effect of BIMU-1 was blocked by the highly specific 5-HT$_4$ antagonist, GR125487.

Several other studies are in favor of a role for 5-HT$_4$ in memory. Following preliminary results using RS 66331,[45] a 5-HT$_4$ agonist/5-HT$_3$ antagonist, the Syntex-Roche group reported that RS 67333, a potent, selective and hydrophobic 5-HT$_4$ agonist, reversed the rat performance deficit induced by atropine in the Morris water maze learning test,[46] an effect reversed by a selective 5-HT$_4$ antagonist RS 67532. In contrast, no effect was seen after injection of RS 67506, a hydrophilic 5-HT$_4$ agonist of equivalent potency and selectivity to RS 67333.[46] This differential effect of RS 67333 and RS 67506 may reflect the enhanced ability of RS 67333 to enter the central nervous system, with respect to RS 67506. None of these substances had any effect on this learning test when injected alone. They were also devoid of any effect on swim speed.[46] Similarly, BIMU-1 and BIMU-8 improve performance after hypoxia-induced amnesia in mice.[47] Pretraining injection of BIMU-1 and BIMU-8 enhanced the acquisition of learning, while the post-training administration impairs the consolidation in the auto-shaping learning task.[48] Finally, Synthelabo Laboratories have a partial 5-HT$_4$ partial agonist under development (SL-650102 for the treatment of Alzheimer's disease (phase I).

Anxiety

Two recent reports indicate that potent and selective 5-HT$_4$ antagonists have weak anxiolytic properties.[49,50] Several tests have been used by Kennett et al to test the anxiolytic properties of

Fig. 3.6. Involvement of 5-HT$_4$ receptors in an olfactory associative task. Mean performance (±SE) obtained across the five 60-trial sessions by rats injected with BIMU-1 (10 mg/kg) (n=7) or saline (control) rats (n=7). The intertrial interval was 15 s. (A) Mean correct response rate to make two odor-reward associations. (B) Mean latencies (in seconds, S$^+$ are the latencies recorded for the positive odor and S$^-$ are the latencies for the negative odor). (C) Reversal test: performance expressed as mean values (±SE) for the control and BIMU-1 rats injected 30 min before the third session one month earlier. On the reversal test valence of the previous learned odors on the five sessions was reversed. The intertrial interval was 15 s. N=7 for each group. On the Y axis are reported the percent of correct responses. Data taken from Marchetti-Gauthier et al.[44]

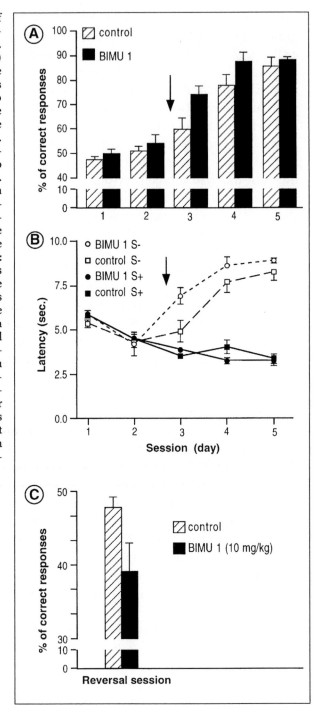

Table 3.2. Anxiolytic properties of 5-HT4 antagonists

Antagonist	Reference	Social Interaction	Elevated Plus Maze		Geller Seifter Conflict Model
			%TOA*	%EOA**	
SB 204070A	Silvestre et al[96]		0	0	
0.3			+	0	
1			0	0	
3					
SB 204070A	Kennett et al[96]				
0.001		+	0	0	ND
0.01		+	+	+	ND
0.1		+	0	0	0
1		0	+	0	0
10		0	ND	ND	ND
SB 207266A					
0.01		+	0	0	ND
0.1		0	0	0	ND
1		+	0	+	0
10		+	0	0	ND

* % time on open arm
** % entries to open arm
ND = not determined

SB204070A and SB207266A.[49] In a social interaction test between two rats unfamiliar to each other placed under bright white light, the two drugs were active for most of the doses tested (Table 3.2). When active, the magnitude of the effect was similar to that seen in response to the benzodiazepine, chlordiazepoxide.[49] In the elevated plus maze test, the anxiolytic effect was less robust, generally observed for few doses tested and only on the percentage of time spent in open arms (TOA) and not on the percentage of entries into open arms compared to total arms entries (EOA), except for one SB204070A and one SB207266A dose (Table 3.2). Using SB204070A, Silvestre et al found similar results.[50] The apparently inconsistent dose relationships may be accounted for by the weakness of the responses. Finally, SB antagonists were inactive on the Geller-Seifter test which measures the capacity to respond to a conflict procedure.[49] The anxiolytic effect of 5-HT$_4$ receptors are consistent with the location of the receptor in the limbic areas of the brain, particularly the hippocampus, lateral septum and medial amygdala which are thought to be involved in the control of anxiety. The above results may suggest a role for $5\text{-}HT_4$ receptors in types of anxiety most closely modeled by the social interaction test, such as generalized anxiety disorder and social phobia. It has also been reported that SDZ 205557 inhibited the anxiolytic profile of diazepam provided SDZ 205557 was administered in the presence of 5-hydroxytryptophan (to raise the serotoninergic tone) and ritanserin (to prevent anxiolysis induced by $5\text{-}HT_2$ receptor stimulation).[51] It is not easy to explain the apparent discrepancy between these latter results and those using SB antagonists. However, both the difference in the experimental protocols used and the low $5\text{-}HT_4/5\text{-}HT_3$ specificity of SDZ 205557 may be implicated.

Other Behavioral Studies

BIMU-1 and BIMU-8 which are, as already stated, mixed $5\text{-}HT_4$ agonists/$5\text{-}HT_3$ antagonists have antinociceptive effects in mice and rats, using hot-plate, abdominal constriction and paw-pressure tests.[47] These effects were prevented by atropine or hemicholinium, but also in rats in which the nucleus basilis magnocellularis (NBM) was destroyed. Naloxone was inactive. This suggests that the integrity of the cholinergic transmission is necessary for the action of BIMU compounds. It is known that cholinomimetics are able to increase pain threshold in both human and animals. Since BIMU-1 and BIMU-8 increase in pain threshold was inhibited by SDZ 205557 and GR125487, it is likely that $5\text{-}HT_4$ rather than $5\text{-}HT_3$ receptors are implicated.[52] Finally, low doses of BIMU-1 and BIMU-8, ineffective when injected by parental route, were highly effective when injected intracerebroventrically, indicating a central effect of the drug.[47]

GR113808, a $5\text{-}HT_4$-specific antagonist, has been found to reduce ethanol intake in alcohol-preferring rats.[53] Interestingly, there is also some indication that $5\text{-}HT_4$ receptors are involved in morphine place conditioning.[54] It is therefore possible that $5\text{-}HT_4$ receptors are involved in brain reward and reinforcement processes, a function compatible with their localization in nucleus accumbens.

Conclusion

Less than 10 years after its first description in the brain, all the tools needed for investigating central $5\text{-}HT_4$ receptor functions are available—labeled ligands, selective agonists and antagonists, gene sequence. Their regional distribution, their control of cAMP production and K^+ channel activity as well as their animal behavioral effects indicate that a role for the 5-HT$_4$ receptor in cognition is possible. Thus, novel therapeutic areas may emerge, including cognitive dysfunction. However, some other aspects of brain function, such as anxiety and rewarding, remain interesting targets for 5-HT$_4$ ligands. The high density of 5-HT$_4$ receptors in the nigro-striatal pathway has not yet been associated with any function related to control of movements, but research in this area is still ongoing.

References

1. Dumuis A, Bouhelal R, Sebben M et al. A non classical 5-hydroxytryptamine receptor positively coupled with adenylate cyclase in the central nervous system. Mol Pharmacol 1988; 34:880-887.
2. Clarke DE, Craig DA, Fozard JR. The 5-HT$_4$ receptor: Naughty but nice. Trends Pharmacol Sci 1989; 10:385-386.
3. Bockaert J, Fozard J, Dumuis A et al. The 5-HT$_4$ receptor: A place in the sun. Trends Pharmacol Sci 1992; 13:141-145.
4. Bockaert J, Ansanay H, Waeber C et al. 5-HT$_4$ receptors. Potential therapeutic implications in neurology and psychiatry. CNS Drugs 1994; 1:6-15.
5. Eglen RM, Wong EHF, Dumuis A et al. Central 5-HT$_4$ receptors. Trends Pharmacol Sci 1995; 16:391-398.
6. Ford APDW, Clarke DE. The 5-HT$_4$ receptor. Med Res Rev 1993; 13:633-662.
7. Andrade R, Chaput Y. 5-hydroxytryptamine$_4$ receptors mediate the slow excitatory response to serotonin in the rat hippocampus. J Pharmacol Exp Ther 1991; 257:930-937.
8. Bockaert J, Sebben M, Dumuis A. Pharmacological characterization of 5-HT$_4$ receptors positively coupled to adenylate cyclase in adult guinea pig hippocampal membranes: Effect of substituted benzamide derivatives. Mol Pharmacol 1990; 37:408-411.
9. Kandel E, Adel T. Neuropeptides, adenylyl cyclase and memory storage. Science 1995; 268:825-826.
10. Grossman CJ, Kilpatrick GJ, Bunce KT. Development of a radioligand binding assay for 5-HT$_4$ receptors in guinea-pig and rat brain. Br J Pharmacol 1993; 109:618-624.
11. Brown AM, Young TL, Patch TL et al. [^{125}I]-SB207710, a potent, selective radioligand for 5-HT$_4$ receptors. Br J Pharmacol 1993; 110:10P.
12. Jakeman LB, To ZP, Eglen RM et al. Quantitative autoradiography of 5-HT$_4$ receptors in brains of three species using two structurally distinct radioligands, [^{3}H]-GR113808 and [^{3}H]-BIMU-1. Neuropharmacology 1994; 33:1027-1038.
13. Dumuis A, Sebben M, Monferini E et al. Azabicycloalkylbenzimidazolone derivatives as a novel class of potent agonists at the 5-HT$_4$ receptor positively coupled to adenylate cyclase in brain. Naunyn-Schmiedeberg's Arch Pharmacol 1991; 343:245-251.

14. Bockaert J, Fagni L, Sebben M et al. Pharmacological characterization of brain 5-HT$_4$ receptors: relationship between the effect of indole, benzamide and azabicycloalkylbenzimidazolone derivatives. In: Fozard JR, Saxena PR, eds. Serotonin: Molecular Biology, Receptors and Functional Effects. Basel: Birkhäuser Verlag, 1991:220-231.

15. Adham N, Gerald C, Schechter L et al. [³H]-5-Hydroxytryptamine labels the agonist high affinity state of the cloned rat 5-HT$_4$ receptor. Eur J Pharmacol 1996; 304:231-235.

16. Bonhaus DW, Berger J, Adham N et al. [³H]RS 57639, a high affinity, selective 5-HT$_4$ receptor partial agonist, specifically labels guinea pig striatal and rat cloned (5-HT$_{4S}$ and 5-HT$_{4L}$) receptors. Neuropharmacology 1997; 36:671-680.

17. Waeber C, Sebben M, Nieoullon A et al. Regional distribution and ontogeny of 5-HT$_4$ binding sites in rodent brain. Neuropharmacology 1994; 33:527-541.

18. Vilaro MT, Cortes R, Gerald C et al. Localization of 5-HT$_4$ receptor mRNA in rat brain by in situ hybridization histochemistry. Mol Brain Res 1996; 43:356-360.

19. Compan V, Daszuta A, Salin P et al. Lesion study of the distribution of serotonin 5-HT$_4$ receptors in rat basal ganglia and hippocampus. Eur J Neurosci 1996; 8:2591-2598.

20. Domenech T, Beleta J, Fernandez AG et al. Identification and characterization of serotonin 5-HT$_4$ receptor binding sites in human brain: comparison with other mammalian species. Mol Brain Res 1994; 21:176-180.

21. Waeber C, Sebben M, Grossman C et al. [³H]-GR113808 labels 5-HT$_4$ receptors in the human and guinea-pig brain. NeuroReport 1993; 4:1239-1242.

22. Schiavi GB, Brunet S, Rizzi CA et al. Identification of serotonin 5-HT$_4$ receptor sites in the porcine caudate nucleus by radioligand binding. Neuropharmacology 1994; 33:543-549.

23. Reynolds GP, Mason SL, Meldrum A et al. 5-Hydroxytryptamine (5-HT$_4$) receptors in post mortem human brain tissue: distribution, pharmacology and effects of neurodegenerative diseases. Br J Pharmacol 1995; 114:993-998.

24. Patel S, Roberts J, Moorman J et al. Localization of serotonin-4 receptors in the striato-nigral pathway in rat brain. Neuroscience 1995; 69:1159-1167.

25. Torres GE, Arfken CL, Andrade R. 5-hydroxytryptamine$_4$ receptors reduce after-hyperpolarization in hippocampus by inhibiting calcium-induced calcium release. Mol Pharmacol 1996; 50:1316-1322.

26. Consolo S, Arnaboldi S, Giorgi S et al. 5-HT$_4$ receptor stimulation facilitates acetylcholine release in rat frontal cortex. NeuroReport 1994; 5:1230-1232.

27. Ullmer C, Engels P, Al Samir A et al. Distribution of 5-HT$_4$ receptor mRNA in the rat brain. Naunyn-Schmiedeberg's Arch Pharmacol 1996; 354:210-212.

28. Boddeke HWGM, Kalkman HO. Zacopride and BRL 24924 induce an increase in EEG-energy in rats. Br J Pharmacol 1990; 101:281-284.

29. Gerald C, Adham N, Kao H-T et al. The 5-HT$_4$ receptor: molecular cloning and pharmacological characterization of two splice variants. EMBO J 1995; 14:2806-2815.

30. Claeysen S, Sebben M, Journot L et al. Cloning, expression and pharmacology of the mouse 5-HT$_{4L}$ receptor. FEBS Letters 1996; 398:19-25.

31. Ansanay H, Sebben M, Bockaert J et al. Pharmacological comparison between [^{3}H]-GR113808 binding sites and functional 5-HT$_4$ receptors in neurons. Eur J Pharmacol 1996; 298:165-174.

32. Ansanay H, Sebben M, Bockaert J et al. Characterization of homologous 5-HT$_4$ receptor desensitization in colliculi neurons. Mol Pharmacol 1992; 42:808-816.

33. Rondé P, Ansanay H, Dumuis A et al. Homologous desensitization of 5-hydroxytryptamine$_4$ receptors in rat oesophagus: functional and second messenger studies. J Pharmacol Exp Ther 1995; 272:977-983.

34. Radja F, Descarries L, Dewar KM et al. Serotonin 5-HT$_1$ and 5-HT$_2$ receptors in adult rat brain after neonatal destruction of nigrostriatal dopamine neurons: a quantitative autoradiographic study. Brain Res 1993; 606:273-285.

35. Karstaedt PJ, Kerasidis H, Pincus JH et al. Unilateral destruction of dopamine pathways increases ipsilateral striatal serotonin turnover in rats. Exp Neurol 1994; 126:25-30.

36. Benloucif S, Keegan MJ, Galloway MP. Serotonin-facilitated dopamine release in vivo: pharmacological characterization. J Pharmacol Exp Ther 1993; 265:373-377.

37. Bonhomme N, De Deurwaërdere P, Le Moal M et al. Evidence for 5-HT$_4$ receptor subtype involvement in the enhancement of striatal dopamine release induced by serotonin: a microdialysis study in the halothane-anesthetized rat. Neuropharmacology 1995; 34:269-279.

38. Ge J, Barnes NM. Further characterization of the 5-HT$_4$ receptor modulating extracellular levels of dopamine in the rat striatum in vivo. Br J Pharmacol 1995; 116:233P.

39. Steward LJ, Ge J, Barnes NM. Ability of 5-HT$_4$ receptor ligands to modify rat striatal dopamine release in vitro and in vivo. Br J Pharmacol 1995; 114:381 P.

40. Bijak M, Tokarski K, Maj J. Repeated treatment with antidepressant drugs induces subsensitivity to the excitatory effect of 5-HT$_4$ receptor activation in the rat hippocampus. Naunyn-Schmiedeberg's Arch Pharmacol 1997; 355:14-19.

41. Ansanay H, Dumuis A, Sebben M et al. A cyclic AMP-dependent, long-lasting inhibition of a K$^+$ current in mammalian neurons. Proc Natl Acad Sci USA 1995; 92:6635-6639.

42. Fagni L, Dumuis A, Sebben M et al. The 5-HT$_4$ receptor subtype inhibits K$^+$ current in colliculi neurons via activation of a cyclic AMP-dependent protein kinase. Br J Pharmacol 1992; 105:973-979.

43. Letty S, Child R, Dumuis A et al. $5-HT_4$ receptors improve social olfactory memory in the rat. Neuropharmacology 1997; 36:681-687.

44. Marchetti-Gauthier E, Roman FS, Dumuis A et al. BIMU1 increases associative memory in rats by activating $5-HT_4$ receptors. Neuropharmacology 1997; 36:697-706.

45. Fontana DJ, Wong EHF, Clark R et al. Pro-cognitive effects of RS-66331, a mixed $5-HT_3$ receptor antagonist / $5-HT_4$ receptor agonist. Proceedings, Third IUPHAR satellite meeting on serotonin, Chicago, July 30th-August 3rd 1994; 57 abstract no. 6.

46. Fontana DJ, Daniels SE, Wong EHF et al. The effects of novel, selective 5-hydroxytryptamine $(5-HT)_4$ receptor ligands in rat spatial navigation. Neuropharmacology 1997; 36:689-696.

47. Ghelardini C, Meoni P, Galeotti N et al. Effect of the two benzimidazolone derivatives: BIMU-1 and BIMU-8 on a model of hypoxia-induced amnesia in the mouse. Proceedings, third IUPHAR satellite meeting on serotonin, Chicago July 30th-August 3rd 1994; P 55.

48. Meneses A, Hong E. Effects of $5-HT_4$ receptor agonists and antagonists in learning. Pharmacol Biochem Behav 1997; 56:347-351.

49. Kennett GA, Bright F, Trail B et al. Anxiolytic-like actions of the selective $5-HT_4$ receptor antagonists SB 204070A and SB 207266A in rats. Neuropharmacology 1997; 36:707-712.

50. Silvestre JS, Fernandes A, Palacios JM. Effects of $5-HT_4$ receptor antagonists on rat behavior in elevated plus-maze. Eur J Pharmacol 1996; 309:219-222.

51. Naylor RJ. A functional role for $5-HT_4$ receptors in the brain? Eur J Neuropsychopharmacol 1993; 3:248-254.

52. Ghelardini C, Galéotti N, Casamenti F et al. Central cholinergic antinociception induced by $5-HT_4$ agonists: BIMU 1 and BIMU 8. Life Sci 1996; 58:2297-2309.

53. Panocka I, Ciccociopo R, Polidori C et al. The $5-HT_4$ receptor antagonist, GR113808, reduces ethanol intake in alcohol-preferring rats. Pharmacol Biochem Behav 1995; 52:255-259.

54. Bisaga A, Sikora J, Kostowski W. The effect of drugs interacting with serotonergic and $5-HT_4$ receptors on morphine place conditioning. Pol J Pharmacol 1993; 45:513-519.

CHAPTER 4

Central 5-HT$_4$ Receptors: Electrophysiology

Rodrigo Andrade, Samir Haj-Dahmane and Esther Chapin

Introduction

Serotonin receptors of the 5-HT$_4$ subtypes are expressed at relatively high densities in several regions of the mammalian central nervous system. As such, these receptors are likely to participate in at least some of the functions served by serotonin in the brain. The challenge is to identify, at a behavioral level, those processes involving 5-HT$_4$ receptors and to elucidate, at cellular and molecular levels, how these processes are made possible by 5-HT$_4$ receptor activation. In this regard, the electrophysiological effects mediated by 5-HT$_4$ receptors constitute an important middle ground. 5-HT$_4$ receptors regulate the electrophysiological properties of individual neurons; thus in turn, the functioning of neuronal networks. It is ultimately the effect of 5-HT$_4$ receptors on the functioning of these networks that impact behavioral and thought processes.

Unfortunately, in 1998, we are far from fulfilling this integrative program. Nevertheless, the past few years have seen steady progress in our understanding of the cellular effects mediated by 5-HT$_4$ receptors. In this chapter, we summarize some of the more salient developments in this area, and try to place them in a larger physiological context. It is hoped that this larger perspective will help in generating functional hypotheses regarding the role of 5-HT$_4$ receptors at the level of neuronal networks and behavior.

5-HT$_4$ Receptors in the Brain and Periphery, edited by Richard M. Eglen.
© 1998 Springer-Verlag and R.G. Landes Company.

5-HT₄ Receptors in Hippocampus

The hippocampus receives a moderately dense serotonergic innervation originating from the median and dorsal raphe nuclei.[1] Early studies demonstrated that the most obvious effect of serotonin in vivo[2] or in vitro[3-6] is a membrane hyperpolarization that resulted in a reduction in membrane excitability. An interesting observation from these early studies was that this inhibitory response, which is mediated by receptors of the 5-HT$_{1A}$ subtype,[7-9] coexisted on the same cells with a second slower "excitatory" response.[8-10]

By blocking the serotonin-induced hyperpolarization using 5-HT$_{1A}$ receptor antagonists, it was possible to study the "excitatory" response to serotonin in isolation. This allowed for a better characterization of this second response and lead to the realization that it was comprised of two components.[8,9] The first of these was a slow membrane depolarization which brought the cell closer to threshold (Fig. 4.1). The second was a reduction in the afterhyperpolarization (AHP), which in these cells, is mediated by a calcium-activated potassium current (Fig. 4.1A). Since this AHP normally limits firing activity in response to excitatory stimuli, its reduction by serotonin greatly enhanced cellular excitability (Fig. 4.1B). However, pharmacological studies could not classify the receptor involved into any of the known subtypes described at the time.[9,11] Hence it represented an "orphan" response.

Shortly after the discovery of this "orphan" serotonin response, the laboratory of Joël Bockaert published a series of seminal biochemical studies demonstrating that the serotonin receptor mediating adenylate cyclase activation in embryonic mouse collicular neurons and in guinea pig hippocampus corresponded to a new serotonin receptor subtype.[12,13] Since at the time only three serotonin receptors had been identified, they named this new subtype 5-HT₄. It was immediately clear that the serotonin receptor mediating the excitatory effects in hippocampus shared some important pharmacological characteristics with this newly discovered 5-HT₄ receptor, including sensitivity to tropisetron (ICS 205-930) and the 4-amino-5-chloro-2 methoxy benzamide derivative renzapride (BRL 24924).[11,14,15] However, one important difference was that renzapride appeared to act as an antagonist in hippocampus, but as a full agonist in the cultured collicular neurons.[11,16] This made it difficult to equate the serotonin receptor mediating the depolarization and the reduction in the AHP with the newly described 5-HT₄ receptor. This

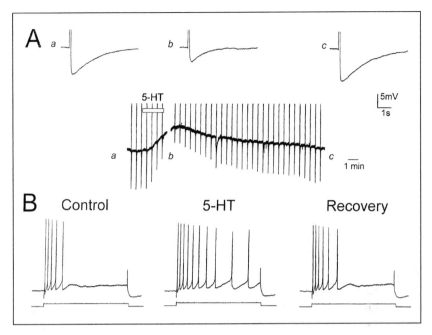

Fig. 4.1. Serotonin depolarizes pyramidal cells of the CA1 region and reduces the AHP. (A) Administration of 5-HT in the presence of a 5-HT$_{1A}$ antagonist results in a slow membrane depolarization. Concurrent with the depolarization there is a decrease in the AHP (downward deflections in the voltage record) that follows a calcium spike (upward deflections in the voltage record). Illustrations a, b, and c correspond to AHPs taken before, during and after serotonin and are displayed using an expanded time base. (B) Response of a different cell to a 600 ms long depolarizing pulse before, during and after serotonin applied in the presence of a 5-HT$_{1A}$ receptor antagonist. Notice that the same pulse is much more effective in the presence of serotonin. Modified from refs. 11,18.

discrepancy could reflect a difference in the receptors involved, or a difference in the coupling of a single receptor to the different effects being measured. Support for the later possibility emerged from the realization that substituted benzamides were partial agonists at both the 5-HT$_4$ receptor coupled to adenylate cyclase activation in guinea pig hippocampus[15] and also at the serotonin receptor mediating the excitatory response to serotonin in rat hippocampus.[17] These results lead to the proposal that the serotonin receptor responsible for the depolarization and the reduction of the AHP in the CA1 region of the hippocampus belonged to the 5-HT$_4$ subtype.[17]

With the development of selective 5-HT$_4$ antagonists it was possible to examine in a more rigorous manner the pharmacology of the serotonin receptor mediating the depolarization and the reduction of the AHP in hippocampus. This was accomplished by using a series of 5-HT$_4$ antagonists (DAU 6285, SDZ 205-930 and GR113808) that varied in affinity for the 5-HT$_4$ receptor over close to a 1,000-fold range.[18] As predicted for a 5-HT$_4$ receptor, all these drugs functioned as antagonists at the 5-HT receptor, mediating the depolarization and the reduction of the AHP (Fig. 4.2) and the rank order of potencies for these responses was that predicted for a 5-HT4 receptor (GR113808> SZ 205-557 > DAU 6285). Furthermore, the apparent potency of these compounds for inhibiting the effect of serotonin on the AHP exhibited an excellent correlation with the affinity of these compounds for the 5-HT$_4$ receptor[18] (Fig. 4.2B). This pharmacological profile persuasively argued for the identification of the serotonin receptor mediating the slow depolarization and the reduction of the AHP in the CA1 region of the hippocampus as belonging to the 5-HT$_4$ subtype.

By identifying a physiological response mediated by 5-HT$_4$ receptors in the brain, it has become possible to begin investigating its physiological regulation. One intriguing feature that has emerged is that the function of 5-HT$_4$ receptors in the CA1 region is modulated by adrenal steroids.[19] Exactly how such modulation might work in vivo, however, it is hard to predict since the effects of corticosteroids on 5-HT$_4$ responses are complex, varying with the time course of treatment. CA1 pyramidal neurons express high levels of both mineralocorticoid and glucocorticoid receptors. Thus, 5-HT$_4$ receptors might represent one of the downstream targets regulated by adrenal steroids in hippocampus.

The advent of high affinity ligands for the 5-HT$_4$ receptor made it possible to examine their localization in the brain. As expected from the physiological results outlined above, 5-HT$_4$ receptors are highly expressed in the CA1 region. Interestingly, a high level of expression was also detected in the other CA field as well as the dentate gyrus.[20,21] This suggests that 5-HT$_4$ receptors might regulate neuronal properties not only in the CA1 region but throughout hippocampus. Indeed, serotonin reduces the AHP in granule cells of the dentate gyrus[22] and pyramidal cells of the CA3 region (S. Beck, personal communication). Although the pharmacology of the receptors involved has not been investigated systematically, these observations suggest 5-HT$_4$ receptors might regulate neuronal activity in these areas as well.

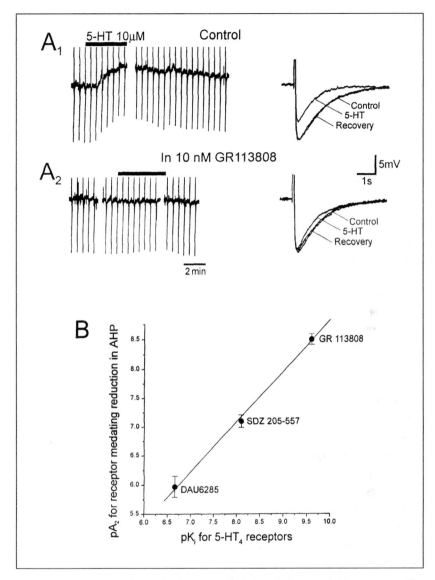

Fig. 4.2. 5-HT₄ receptors mediate the serotonin-induced depolarization and reduction of the AHP. (A) Administration of 10 nM GR 113808 blocks the serotonin induced depolarization and reduction of the AHP recorded in isolation following blockade of 5-HT₁ₐ receptors in a CA1 pyramidal neuron. (B) Correlation between the apparent potency (pA₂) of DAU 6285, SDZ 205-557 and GR 113808 at the serotonin receptor mediating the reduction of the AHP in the CA1 region and at 5-HT₄ receptors identified in biochemical assays. Modified from ref. 18.

Regulation of the AHP

Given the ability of 5-HT_4 receptors to stimulate adenylate cyclase,[13] an obvious possibility is that these receptors reduce the AHP through a cAMP dependent mechanism. Consistent with this possibility, it has been known for many years that increases in intracellular cAMP in pyramidal cells of the CA1 region result in a reduction of the AHP[23,24] (Fig. 4.3A). However, such mimicry is not sufficient to establish a role for cAMP in mediating the effects of 5-HT_4 receptors, since it is known that other neurotransmitters can reduce the AHP in this region by acting through cAMP independent mechanisms.[25]

In many systems a standard strategy to establish a role for cAMP has been to use phosphodiesterase inhibitors. Since phosphodiesterases hydrolyze cAMP, thus terminating its action, an enhancement of a response by these drugs would support the involvement of cyclic nucleotides. However, in the case of the 5-HT_4 mediated reduction of the AHP, phosphodiesterase inhibitors produced ambiguous[9] or relatively small enhancement[26] of the 5-HT_4 responses in the CA1 region. Thus an alternative approach was needed to provide compelling evidence for, or against, the involvement of cAMP.

A second approach to this problem utilizes the fact that many cAMP mediated responses involve stimulation of protein kinase A (PKA). Thus, the involvement of PKA would strongly argue for a role for cAMP. Since there are several inhibitors for this kinase, this provides a second avenue for testing the possible involvement of cAMP. This approach was used in hippocampus. Consistent with the involvement of cAMP in the 5-HT_4 induced reduction of the AHP in the CA1 region, staurosporine reduced the ability of serotonin to inhibit the AHP in a concentration dependent manner.[26] Staurosporine is a broad spectrum protein kinase inhibitor that differs markedly in its potency for different kinases.[27] Since staurosporine inhibited the 5-HT_4 receptor mediated effect on the AHP with a potency comparable to that for a known PKA mediated response,[26] these results supported the possible involvement of PKA.

Additional evidence for the involvement of cAMP and PKA in the 5-HT_4 receptor mediated inhibition of the AHP comes from the use of the cAMP analog (Rp)-cAMPs. This compound competes with cAMP for the regulatory site of PKA, but does not activate the kinase.[28] Injection of (Rp)-cAMPs into pyramidal cells of the CA1 region resulted in a marked inhibition in the ability of 5-HT_4

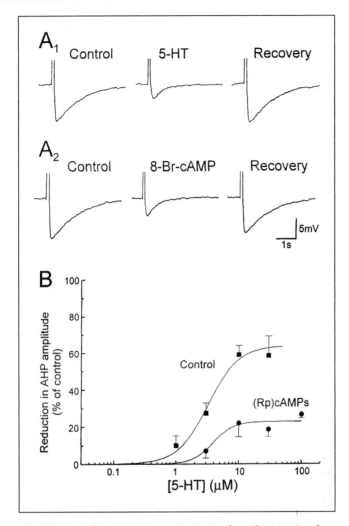

Fig. 4.3.The ability of 5-HT$_4$ receptors to reduce the AHP involves cAMP. (A$_1$) Effect of serotonin on the AHP in the presence of a 5-HT$_{1A}$ antagonist on a CA1 pyramidal neuron. (A$_2$) Effect of 1 mM 8-bromo cAMP in the same neuron. (B) Concentration relation relationship for the ability of serotonin to inhibit the AHP under control conditions and following intracellular injection of (Rp)cAMPs. Modified from ref. 26.

receptors to suppress the AHP[26] (Fig. 4.3B). This treatment also inhibited β-adrenergic responses, which are mediated by PKA, but not other PKA-independent responses in these same cells. These results strongly argued for the involvement of cAMP and PKA in mediating 5-HT$_4$ responses in the CA1 region. Similar effects with staurosporine, (Rp)-cAMPs and PKI, a highly selective peptide inhibitor of PKA, were also obtained in CA1 neurons in brain slices from young rats.[29] Although this last study did not use pharmacological means to identify or isolate 5-HT$_4$ responses, it seems likely that at least a portion of the serotonin-induced reduction of the AHP was mediated by 5-HT$_4$ receptors. As such, these results again support the involvement of cAMP and PKA in this response.

A recent study has examined the mechanisms downstream from PKA that mediate the 5-HT$_4$-receptor mediated reduction of the AHP.[30] Initial studies had shown that the 5-HT$_4$ receptor mediated reduction of the AHP was not mediated by a reduction in calcium influx into the cell.[9] This suggested that the site of action for PKA was downstream from the voltage activated calcium channels whose opening triggers the AHP. One intriguing possibility was raised by the observation that in several neuronal cell types,[31,32] including pyramidal cells of the CA1 region,[30] calcium-induced calcium release from intracellular calcium stores plays an important role in the generation of the AHP. Since the intracellular calcium channel responsible for calcium-induced calcium release is a target for PKA phosphorylation,[33] this raised the possibility that 5-HT$_4$ receptors could target calcium-induced calcium release to reduce the AHP. Consistent with this possibility, the ability of 5-HT$_4$ receptors to reduce the AHP was significantly increased when the contribution from intracellular calcium release to the AHP was enhanced. Furthermore, following inactivation of calcium-induced calcium release, either through blockade of the calcium release channel or depletion of the intracellular calcium stores, 5-HT$_4$ receptor activation failed to reduce the AHP. Combined, these results lead to the conclusion that 5-HT$_4$ receptors reduce the AHP in the CA1 region in the hippocampus by stimulating cAMP, activating PKA and reducing calcium-induced calcium release.

Calcium-induced calcium release is an ubiquitous mechanism that functions to enhance calcium transients in a variety of cell types. Since calcium transients play an ubiquitous role in cellular signaling, the ability of $5\text{-}HT_4$ receptors to regulate this process suggests they might not only regulate cellular excitability, but also play a role in regulating calcium signaling in central neurons.

$5\text{-}HT_4$-Receptor-Mediated Slow Depolarization

While the $5\text{-}HT_4$ mediated reduction in the AHP has received the most attention, activation of $5\text{-}HT_4$ receptors in the CA1 region of the hippocampus also elicits a slow membrane depolarization. Unfortunately, the mechanism for this depolarization is not well understood at present. In thalamic neurons,[34] neurons of the nucleus prepositus hypoglossi[35] and motoneurons of the spinal cord,[36] serotonin has been shown to elicit a slow membrane depolarization that superficially resembles that seen in hippocampus. This effect is signaled through cAMP and is mediated by a voltage dependent cation nonselective current generally referred to as I_h. Given the similarity in the responses, an obvious possibility is that $5\text{-}HT_4$ receptors depolarize pyramidal cells of the CA1 through similar mechanisms. Surprisingly, several distinct lines of evidence suggest this is probably not the case. The $5\text{-}HT_4$-induced depolarization in the CA1 region, unlike that seen, for example in the thalamus, is associated with a conductance decrease.[9] Furthermore, the depolarization in hippocampus persists unaltered following the blockade of I_h.[37] Finally, the amplitude of the $5\text{-}HT_4$ depolarization diminishes with hyperpolarization, a result contrary to what would be expected for a depolarization mediated by I_h. Since all of these properties are inconsistent with a mediation by I_h it seems likely that the $5\text{-}HT_4$ depolarization is mediated by an alternative mechanism, possibly a reduction in a resting potassium conductance.

Interestingly, the pharmacology of the serotonin receptors mediating the enhancement of I_h suggests that they cannot be classified in the $5\text{-}HT_4$ subfamily of serotonin receptors.[35-36] This raises the intriguing possibility that perhaps different G_s coupled serotonin receptors might also couple to different ion channels in central neurons. Future studies will be needed to examine these possibilities.

$5\text{-}HT_4$ Receptors in Cultured Embryonic Collicular Neurons

Mouse embryonic collicular neurons express $5\text{-}HT_4$ receptors that couple to adenylate cyclase.[12,13] This is the classical preparation where $5\text{-}HT_4$ receptors were initially identified and several studies from the Bockaert laboratory have examined the effects mediated by 5-HT$_4$ receptors in these cells. One surprising finding from these studies is that these cells do not seem to express 5-HT$_4$ receptors in situ,[20] although they do so in primary culture. These caveat notwithstanding, these cells remain a valuable model system in which to study the electrophysiology of 5-HT$_4$ receptors.

Administration of 5-HT to embryonic collicular neurons in primary culture results in a reduction of the AHP and a decrease in spike frequency accommodation similar to that seen in the CA1 region of hippocampus.[38] In addition, there is also an increase in the duration of the action potential.[38] Since these effects were blocked by DAU 6285, they appear to be mediated by receptors of the 5-HT$_4$ subtype. Furthermore, since these effects were mimicked by administration of the membrane permeant cAMP analog 8-bromo cAMP, they also appeared to be mediated by stimulation of adenylate cyclase.

While the precise mechanism underlying each of these macroscopic effects on cellular excitability have not been elucidated, voltage clamp recordings have identified a potassium current that is inhibited by 5-HT in these cells.[38,39] This effect of serotonin is mimicked by substituted benzamides such as renzapride, and blocked by DAU 6285, indicating that it is mediated by the activation of 5-HT$_4$ receptor.[39] The effect of 5-HT$_4$ receptor activation on these cells is also mimicked by the administration of cholera toxin and 8-bromo cAMP, and is inhibited by H7 and PKI.[38,39] This indicates that 5-HT$_4$ receptors in these cells signal their effect on the potassium current by stimulating cAMP formation and activating PKA.

One interesting feature of the potassium current inhibited by 5-HT$_4$ receptors is that it is active only at depolarized potentials (above -40 mV) and is inhibited by tetraethylammonium and 4-aminopyridine. These characteristics suggest that it belongs to the family of voltage-sensitive potassium channels. As such, this current is probably different from those regulated by 5-HT$_4$ receptors in hippocampus (see above). Nevertheless, the properties of this current are such that it could contribute to the macroscopic effects of serotonin in these cells, particularly the increase in action potential duration.

Little is presently known about the functional significance of inhibiting this current. However, since increases in action potential duration can result in the enhancement of neurotransmitter release, one possible physiological role for the 5-HT_4 inhibition of this current could be presynaptic facilitation.[38] Consistent with this hypothesis, recent studies indicate that at least some 5-HT_4 receptors can be targeted to presynaptic terminals.[40] If this hypothesis is correct, an intriguing feature of this 5-HT_4 response could prove to be functionally important. In these cells the 5-HT_4 receptor induced inhibition of the potassium channel seems to involve not only activation of PKA, but also inactivation of a phosphatase.[38] As a result, even transient 5-HT_4 receptor activation results in prolonged, up to 2 hours, inhibition of the potassium current. Thus, 5-HT_4 receptors might be able to affect relatively long lasting changes in synaptic efficacy and hence could participate in some forms of synaptic plasticity.[38]

Conclusions

From the studies outlined above, it is clear that the electrophysiology of 5-HT_4 receptors is still in its infancy. While these receptors are widely distributed in the mammalian central nervous system, our appreciation of their possible function in modulating neuronal excitability and function is restricted to only a very limited number of cell types. As such, it is likely that future studies will add much to our understanding of the mechanism by which 5-HT_4 receptors can control the functioning of neurons and neuronal networks.

In spite of these limitations, a few general features have begun to emerge. The simplest is that 5-HT_4 receptors appear to target, either directly or indirectly, potassium channels to increase membrane excitability. This pattern is seen in hippocampal pyramidal cells and collicular neurons and probably holds for many other cell types as well. Cyclic AMP and PKA play a role in these actions, but it is possible that future studies might reveal PKA independent, or perhaps even cAMP independent responses mediated by 5-HT_4 receptors. Furthermore, given our understanding of the range of responses signaled through cAMP, it is likely that additional effects of 5-HT_4 receptors beyond potassium channels will be uncovered. In this regard, the ability of 5-HT_4 receptors to regulate calcium-induced calcium release in hippocampus might represent just a hint of things to come.

A second feature is that most neurons in the brain probably do not express 5-HT_4 receptors in isolation. This is supported not only by the electrophysiological results in hippocampus, but also by the extensive colocalization of 5-HT_4 receptors with other serotonin receptor subtypes in the brain. Thus, 5-HT_4 receptors in a given area, and in a given cell type, are not likely to be the sole mediators of serotonin actions. More likely, 5-HT_4 receptors will be found to act in concert with other receptor subtypes to mediate the effects of serotonin, even at the single cell level. This is likely to be important for understanding the role played by 5-HT_4 receptors in regulating neuronal function. For example, in hippocampus, 5-HT_{1A}, 5-HT_4 and other receptor subtypes are coexpressed in pyramidal cells of the CA1 region. While 5-HT_4 receptors activated in isolation seem to be excitatory, their effect when coactivated with 5-HT_{1A} receptors is considerably more subtle. In this case serotonin neither excites nor inhibits the cells, but instead alters the manner in which they integrate incoming information and transform it into firing output.[9] This has important consequences for the ways in which we think about 5-HT_4 receptors. For example, a reduction in 5-HT_4 receptor function in this area will not result in a reduction in the excitatory effect of serotonin, but more likely will produce a shift in the way cells process information in response to serotonin. If these hippocampal results are generalizable, then to understand the actions of 5-HT_4 receptors, we will have to consider not only what 5-HT_4 receptors do, but also how these actions mesh with those of other receptors at the cellular level.

Acknowledgments

      Work in the authors' laboratory is supported by grants MH 43985 and MH49355 from the National Institutes of Health (USA).

References

1. Azmitia EC, Segal M. An autoradiographic analysis of the differential ascending projections of the dorsal and median raphe nuclei in the rat. J Comp Neurol 1978; 179:641-668.
2. Biscoe TJ, Straughan DW. Micro-electrophoretic studies of neurones in the cat hippocampus. J Physiol (London) 1966; 183:341-359.
3. Segal M. The action of serotonin in the rat hippocampal slice preparation. J Physiol (London) 1980; 303:423-439.
4. Jahnsen H. The action of 5-hydroxytryptamine on neuronal membranes and synaptic transmission in area CA1 of the hippocampus in vitro. Brain Res 1980; 197:83-94.
5. Andrade R, Malenka RC, Nicoll RA. A G protein couples serotonin and GABAB receptors to the same channels in hippocampus. Science 1986; 234:1261-1265.
6. Beck SG, Goldfarb J. Serotonin produces a reversible concentration dependent decrease of population spikes in rat hippocampal slices. Life Sci 1985; 36:557-563.

7. Beck SG, Clarke WP, Goldfarb J. Spiperone differentiates multiple 5-hydroxytryptamine responses in rat hippocampal slices in vitro. Eur J Pharmacol 1985; 116:195-197.

8. Colino A, Halliwell JV. Differential modulation of three separate K-conductances in hippocampal CA1 neurons by serotonin. Nature 1987; 328:73-77.

9. Andrade R, Nicoll RA. Pharmacologically distinct actions of serotonin on single pyramidal neurones of the rat hippocampus recorded in vitro. J Physiol (London) 1987; 394:99-124.

10. Beck SG. 5-Carboxyamidotryptamine mimics only the 5-HT elicited hyperpolarization of hippocampal pyramidal cells via 5-HT1A receptor. Neurosci Lett 1989; 99:101-106.

11. Chaput Y, Araneda RC, Andrade R. Pharmacological and functional analysis of a novel serotonin receptor in the rat hippocampus. Eur J Pharmacol 1990; 182:441-456.

12. Dumuis A, Bouhelal R, Sebben M et al. A 5-HT receptor in the central nervous system, positively coupled with adenylate cyclase, is antagonized by ICS 205 930. Eur J Pharmacol 1988; 146:187-188.

13. Dumuis A, Bouhelal R, Sebben M et al. A nonclassical 5-Hydroxytryptamine receptor positively coupled with adenylate cyclase in the central nervous system. Mol Pharmacol 1988; 34:880-887.

14. Dumuis A, Sebben M, Bockaert J. BRL 24924: a potent agonist at a non-classical 5-HT receptor positively coupled with adenylate cyclase in colliculi neurons. Eur J Pharmacol 1989; 162:381-384.

15. Bockaert J, Sebben M, Dumuis A. Pharmacological characterization of 5-hydroxytryptamine$_4$ (5-HT$_4$) receptors positively coupled to adenylate cyclase in adult guinea pig hippocampal membranes: effect of substituted benzamide derivatives. Mol Pharmacol 1990; 37:408-411.

16. Dumuis A, Sebben M, Bockaert J. The gastrointestinal prokinetic benzamide derivatives are agonists at the non-classical 5-HT receptor (5-HT$_4$) positively coupled to adenylate cyclase in neurons. Naunyn-Schmiedeberg's Arch Pharmacol 1989; 340:403-410.

17. Andrade R, Chaput Y. 5-HT$_4$-like receptors mediate the slow excitatory response to serotonin in the rat hippocampus. J Pharmacol Exp Ther 1991; 257:930-7.

18. Torres GE, Holt IL, Andrade R. Antagonists of 5-HT$_4$ receptor-mediated responses in adult hippocampal neurons. J Pharmacol Exp Ther 1994; 271(1):255-61.

19. Birnstiel S, Beck SG. Modulation of the 5-hydroxytryptamine$_4$ receptor mediated response by short-term and long-term administration of corticosterone in rat CA1 hippocampal pyramidal neurons. J Pharmacol Exp Ther 1995; 273:1132-8.

20. Waeber C, Sebben M, Nieoullo, A et al. Regional distribution and ontogeny of 5-HT$_4$ binding sits in rodent brain. Neuropharmacology 1994; 33:527-41.

21. Ullmer C, Engels P, Abdel'Al, S et al. Distribution of 5-HT$_4$ receptor mRNA in the rat rain. Naunyn-Schmiedebergs Arch Pharmacol 1996; 354:210-2.

22. Baskys A, Niesen CE, Carlen PL. Altered modulatory actions of serotonin on dentate granule cells of aged rats. Brain Res 1987; 419:112-8.

23. Haas HL, Konnerth A. Histamine and noradrenaline decrease calcium-activated potassium conductance in hippocampal pyramidal cells. Nature 1983; 302:432

24. Madison DV, Nicoll RA. Noradrenaline blocks accommodation of pyramidal cell discharge in the hippocampus. Nature 1982; 299(14): 636-8.

25. Nicoll RA, Malenka RC, Kauer JA. Functional comparison of neurotransmitter receptor subtypes in mammalian central nervous system. Physiol Rev 1990; 70:513-65.

26. Torres GE, Chaput Y, Andrade R. cAMP and PKA mediate 5-HT₄ receptor regulation of calcium-activated potassium current in adult hippocampal neurons. Mol Pharmacol 1995; 47:191-7.

27. Meggio F, Donella Deana A., Ruzzene M et al. Different susceptibility of protein kinases to staurosporin inhibition. Kinetic studies and molecular bases for the resistance of protein kinase CK2. Eur J Biochem 1995; 234:317-22.

28. Parker-Bothelho LH, Rothermel JD, Coombs, RV et al. cAMP analog antagonists of cAMP action. Meth Enzymol 1988; 159:159-72.

29. Pedarzani P, Storm JF. PKA mediates the effects of monoamine neurotransmitters on the K⁺ current underlying the slow spike frequency adaptation in hippocampal neurons. Neuron 1993; 11:1023-35.

30. Torres GE, Arfken CL, Andrade R. 5-hydroxytryptamine₄ receptors reduce afterhyperpolarization in hippocampus by inhibiting calcium-induced calcium release. Mol Pharmacol 1996; (in press).

31. Sah P, McLachlan EM. Ca²⁺-activated K⁺ currents underlying the afterhyperpolarization in guinea pig vagal neurons: A role for Ca²⁺-activated Ca²⁺ release. Neuron 1991; 7:257-64.

32. Kuba K, Morita K, Nohmi M. Origin of calcium ions involved in the generation of a slow afterhyperpolarization in bullfrog sympathetic neurones. Pflugers Archiv 1983; 399:194-202.

33. Furuichi T, Kohda K, Miyawaki A et al. Intracellular channels. Curr Opin Neurobiol 1994; 4:294-303.

34. Pape H, McCormick DA. Noradrenaline and serotonin selectively modulate thalamic burst firing by enhancing a hyperpolarization-activated cation current. Nature 1989; 340:715-8.

35. Bobker DH, Williams JT. Serotonin augments the cationic current Ihin central neurons. Neuron 1989; 2:1535-1450,

36. Takahashi T, Berger AJ. Direct excitation of rat spinal motoneurones by serotonin. J Physiol (London) 1990; 423:63-76.

37. Andrade R, Haj-Dahmane S, Chapin EM. On the mechanim underlying the 5-HT4 receptor-induced depolarization in rat hippocampus. [Abstract] Soc Neurosci Abstr 1997; (in press).

38. Ansanay H, Dumuis A, Sebben M et al. cAMP-dependent long lasting inhibition of a K⁺ current in mammalian neuronses. Proc Natl Acad Sci USA 1995; 92:6635-9.

39. Fagni L, Dumuis A, Sebben M et al. The 5-HT$_4$ receptor subtype inhibits K$^+$ current in colliculi neurones via activation of a cyclic AMP-dependent protein kinase. Eur J Pharmacol 1992; 105:973-9.

40. Patel S, Roberts J, Moorman J et al. Localization of serotonin-4 receptors in the striatonigral pathway in rat bain. Neurosci 1995; 69:1159-67.

CHAPTER 5

Neurochemical Consequences Following Pharmacological Manipulation of Central 5-HT$_4$ Receptors

Janine M. Barnes and Nicholas M. Barnes

Introduction

The monoamine 5-hydroxytryptamine (5-HT; serotonin) functions as both a neurotransmitter and a local hormone via at least 14 structurally and pharmacologically distinct mammalian 5-HT receptor subtypes (for reviews, see refs. 1, 2). This group of receptors possesses members of the putative seven transmembrane domain G-protein-coupled metabotropic receptor family and a member of the ligand-gated ion channel receptor family comparable to receptor groups responsive to glutamate, GABA and acetylcholine. The present chapter focuses on the neurochemical functions of central 5-HT$_4$ receptors.

Identification of the 5-HT$_4$ Receptor

Neurochemical investigations were responsible for the initial identification of the 5-HT$_4$ receptor. Thus, while 5-HT receptors positively coupled to adenylate cyclase had been identified in the 1970s,[3-6] it was not until the end of the 1980s that one of these 5-HT-sensitive receptors was designated as the 5-HT$_4$ receptor by Bockaert and co-workers.[7-10] This 5-HT$_4$ receptor corresponded to the 5-HT receptor enhancing adenylate cyclase activity in guinea pig hippocampal membranes, for which 5-CT (5-carboxamidotryptamine) displayed relatively low potency.[11] Similar 5-HT$_4$ receptor-mediated activation

5-HT$_4$ Receptors in the Brain and Periphery, edited by Richard M. Eglen.
© 1998 Springer-Verlag and R.G. Landes Company.

of adenylate cyclase responses were subsequently demonstrated in various peripheral and central tissue preparations (see ref. 12 for review), including human brain tissue.[13]

While the 5-HT$_4$ receptor was the last 5-HT receptor to be identified without the aid of molecular biological techniques, the subsequent cloning of the cDNA sequence encoding the 5-HT$_4$ receptor confirmed the predicted structure, i.e., the deduced polypeptide sequence displayed the seven putative transmembrane domain topology of G-protein coupled receptors, consistent with the known transduction system associated with the native receptor. The cloning studies also identified the presence of two alternatively spliced variants of the 5-HT$_4$ receptor, both of which couple positively to adenylate cyclase in heterologous expression systems.[14-16] To date, no pharmacological differences between the spliced variants of the 5-HT$_4$ receptor have been reported, and given that a recent report documents only minor differences in their relative regional expression,[17] potential functional differences attributed to the spliced variants remain speculative.

It is of potential interest, however, that since the 5-HT$_4$ receptor spliced variants differ in their carboxy termini,[15] which is known to be involved in the receptor coupling to the G-protein and modification in receptor function following phosphorylation (e.g., desensitization),[18-20] small differences in the efficiency with which the spliced variants couple to adenylate cyclase have been reported (although this may be a consequence of the dissimilarities in the expression levels of the two spliced variants in this study[15]) as have differences in receptor desensitization characteristics in different preparations (see ref. 12).

Modulation of Central Neurotransmitter Release Following Interaction with the 5-HT$_4$ Receptor

Elegant studies have demonstrated that the 5-HT$_4$ receptor-mediated increase in cAMP levels lead to the phosphorylation of a range of target proteins by, for instance, cAMP-dependent protein kinase (e.g., voltage-gated K$^+$ channels).[21] Hence, for example, following activation of the neuronally located 5-HT$_4$ receptor, increased neuronal excitability and slowing of repolarization can be detected electrophysiologically (e.g., see refs. 22-24). Consistent with these electrophysiological modifications following 5-HT$_4$ receptor activation, there are now numerous reports demonstrating the ability of the 5-HT$_4$ receptor to modulate neurotransmitter release in the CNS. Initially the modulation of central acetylcholine release via the 5-HT$_4$ receptor received most attention, probably resulting from the well documented ability of the 5-HT$_4$ receptor to facilitate acetylcholine release in the gastrointestinal tract (e.g., refs. 25-30).

5-HT$_4$ Receptor-Mediated Modulation of Central Acetylcholine Release

Consolo and colleagues[31] reported an increase in acetylcholine release in the rat frontal cortex following i.c.v. administration of the 5-HT$_4$ receptor agonists, BIMU1 and BIMU8 (assessed using the microdialysis technique with freely moving animals; Fig. 5.1). While the authors demonstrated a concentration related response with both agonists, the dose response curves were very steep. Also, it is noteworthy that a dose of BIMU1 (100 nmol i.c.v.), which was less than twice the maximal effective dose (60 nmol i.c.v.) and just 10-fold higher than the minimum effective dose (10 nmol i.c.v.), induced only a modest increase in acetylcholine release (55% increase compared to the 147% increase evoked by 60 nmol BIMU1). The pharmacological mechanism underlying the reduced effectiveness of higher concentrations of BIMU1 have not been directly investigated, however, this compound displays appreciable affinity for the 5-HT$_3$ receptor. Pharmacological interpretation is complicated further since BIMU1 (and BIMU8), at higher concentrations interact with another receptor (possibly muscarinic receptors[32]). In addition, BIMU1 behaves as a partial agonist at the 5-HT$_4$ receptor in various preparations (e.g., guinea pig myenteric plexus, mouse embryonic colliculi neurones, piglet atria, rat esophagus; refs. 30, 32-34) and since the level of 5-HT$_4$ receptor reserve differs between tissues (see ref. 35 for review), and the level of receptor reserve in relation to the modulation of central acetylcholine release is unknown, the precise pharmacological action of BIMU1 in the presence of endogenous agonist is difficult to predict.

Despite the difficulty of describing all the actions of BIMU1, there is strong evidence that the BIMU1-induced increase in cortical acetylcholine release was 5-HT$_4$ receptor mediated since the selective 5-HT$_4$ receptor antagonists, GR113808 and GR125487, attenuated the response[31] (Fig. 5.1); GR125487 appeared to be more effective than GR113808, which is consistent with the known higher affinity and longer duration of action of GR125487 relative to GR113808.[36-38] Since

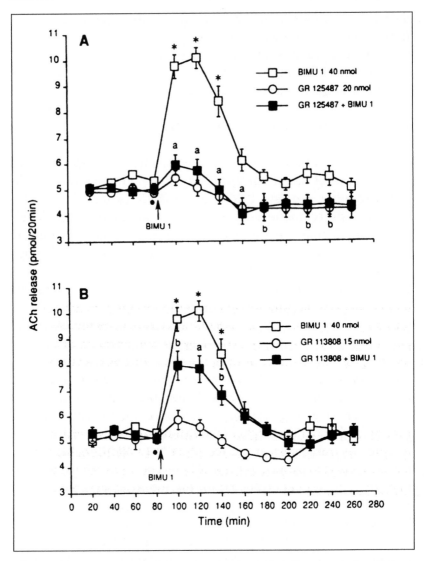

Fig. 5.1. Modulation of acetylcholine release in the rat frontal cortex via the 5-HT₄ receptor, in vivo. The 5-HT₄ receptor agonist BIMU1 (administered i.c.v.; arrow) increased the release of acetylcholine in the rat frontal cortex, assessed using the microdialysis technique, and the BIMU1-induced response was antagonized by administration (i.c.v. 5 min prior to BIMU1; •) of either of the selective 5-HT₄ receptor antagonists GR125487 (A) or GR113808 (B). Data are expressed as pmol of acetylcholine released/20 min (mean ±SEM, n= 6-8). *P<0.01 versus the respective baseline, Dunnett's test. Interactions: [GR125487 + BIMU1] vs. [BIMU1]; [GR113808 + BIMU1] vs. [BIMU1], [a]P<0.01 and [b]P<0.05 by two-way ANOVA followed by Tukey's test for unconfounded means. Reproduced with permission from Consolo et al, Neuroreport 1994; 5:1230-1232.

neither 5-HT$_4$ receptor antagonist alone modified the release of ace-tylcholine[31] (Fig. 5.1), this indicates a lack of endogenous tone on the 5-HT$_4$ receptor in this preparation.

Interestingly, in comparable experiments to those demonstrating a 5-HT$_4$ receptor mediated modulation of cortical acetylcholine release, neither BIMU1 nor BIMU8 modified acetylcholine release in the striatum,[31] despite evidence that cholinergic neurones in these areas may express 5-HT$_4$ receptors. Thus the presence of both 5-HT$_4$ receptor mRNA and binding sites in the striatum, the latter being sensitive to intrastriatal kainic acid lesion,[39,40] suggests that the striatal 5-HT$_4$ receptors are expressed by neurones which have their cell bodies in the striatum, for example striatal cholinergic interneurones (although phenotypically different neurones, e.g., striatal GABA projection neurones, are also candidates).

With respect to the ability of 5-HT$_4$ receptors to modulate hippocampal cholinergic activity, central activation of 5-HT$_4$ receptors following i.c.v. drug administration, increases total EEG-energy, including the theta rhythm arising from the cholinergic septal-hippocampal pathway.[41,42] However, in the same study by Consolo and colleagues[31] that demonstrated the ability of 5-HT$_4$ receptors to modulate cortical, but not striatal, acetylcholine release, they also reported that the 5-HT$_4$ receptor agonists BIMU1 and BIMU8 failed to modulate hippocampal acetylcholine release. The different methodological approaches estimating hippocampal acetylcholine release may underlie this apparent discrepancy. Furthermore, the EEG studies were performed before reports of selective 5-HT$_4$ receptor ligands, and given that some pharmacological differences were evident in the in vivo preparation when compared to studies performed in vitro (e.g., lack of stereoselectivity for the isomers of zacopride to increase EEG energy[42]), re-evaluation of the in vivo response using more selective ligands is warranted. However, the septum (which contains cholinergic cell bodies which project to the hippocampus) expresses moderate levels of 5-HT$_4$ receptor mRNA[43] (although not noted in the text but clearly evident from the autoradiograms) and the hippocampus expresses relatively high levels of 5-HT$_4$ receptors—although these are unlikely to be expressed solely on cholinergic terminals since the hippocampus possesses cells which express 5-HT$_4$ receptor mRNA[17,43,44] and also hippocampal pyramidal neurones express functional 5-HT$_4$ receptors (for example see ref. 24).

Given the well known association of acetylcholine and memory (e.g., ref. 45) and reports that the $5\text{-}HT_4$ receptor facilitates cholinergic function within relevant regions of the brain (e.g., cerebral cortex, hippocampus; see refs. 31, 42), this mechanism provides a plausible explanation for the growing number of reports that demonstrate a facilitation of cognitive performance following $5\text{-}HT_4$ receptor activation[46-48] (but see also ref. 49). However, the likely (additional?) expression of $5\text{-}HT_4$ receptors by noncholinergic neurones in the hippocampus and cerebral cortex (as discussed above) suggests that additional mechanisms may also provide rational explanations for the association of $5\text{-}HT_4$ receptor activation with a facilitation in cognitive performance. For instance, the $5\text{-}HT_4$ receptor mediated excitation of hippocampal pyramidal neurones is likely to facilitate the induction of LTP which is widely regarded as a cellular basis of memory (e.g., ref. 50).

$5\text{-}HT_4$ Receptor-Mediated Modulation of Central Dopamine Release

There is now compelling neurochemical evidence that the $5\text{-}HT_4$ receptor modulates striatal dopamine release. Initially, Benloucif and colleagues[51] implicated a role for the $5\text{-}HT_4$ receptor in the 5-HT modulation of striatal dopamine release by demonstrating that the nonselective $5\text{-}HT_4$ receptor agonists, 5-HT and 5-methoxytryptamine, increased striatal dopamine release in anesthetized rats and the response was partially blocked by high concentrations of the weak nonselective $5\text{-}HT_4$ receptor antagonist, tropisetron. However, use of the natural agonist 5-HT (and possibly the structurally related agonist 5-methoxytryptamine) to evoke dopamine release complicates subsequent pharmacological analysis of receptor mediation since 5-HT also acts as a substrate for the dopamine transporter, resulting in an elevation of extracellular levels of dopamine (e.g., ref. 52). Hence in the absence of dopamine reuptake blockers, receptor antagonists will not be able to completely block the 5-HT-induced increase in apparent dopamine release, estimated using the microdialysis technique (e.g., refs. 51, 52). Subsequent studies, however, used a range of structurally diverse $5\text{-}HT_4$ receptor ligands, including highly selective $5\text{-}HT_4$ receptor antagonists (e.g., GR118303 and GR125487[36-38]), to confirm that the $5\text{-}HT_4$ receptor enhances the

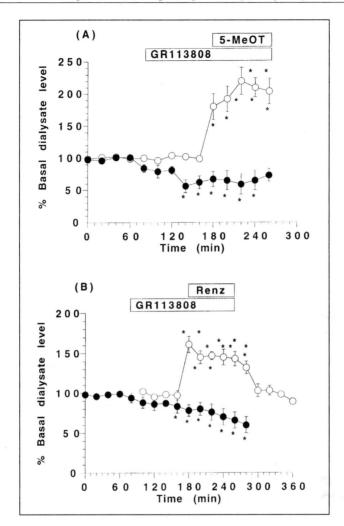

Fig. 5.2. Modulation of dopamine release in the rat striatum via the 5-HT$_4$ receptor, in vivo. The 5-HT$_4$ receptor agonists 5-methoxytryptamine (5-MeOT) and renzapride (Renz) increased the release of dopamine, assessed using the in vivo microdialysis technique, and the agonist-induced response was antagonized by the selective 5-HT$_4$ receptor antagonist GR113808. (A) 5-methoxytryptamine (O; 10 µM; in the presence of pindolol (10 µM) and methysergide (10 µM)) and 5-MeOT (10 µM; in the presence of pindolol (1 µM) and methysergide (1 µM)) plus GR113808 (●; 1 µM) and (B) renzapride (O; 100 µM; Renz) and renzapride (100 µM) plus GR113808 (●; 1 µM). Dopamine levels in dialysates are expressed as the percentage of the mean absolute amount in the 4 collections proceeding the drug treatment. Data represents mean ± SEM, n = 4-6. Horizontal bars represent application of the indicated drug via the perfusing aCSF. ANOVA<0.05, *P<0.05, **P<0.01 (Dunnett's t test). Reproduced with permission from Steward LJ et al, Br J Pharmacol 1996; 117:55-62.

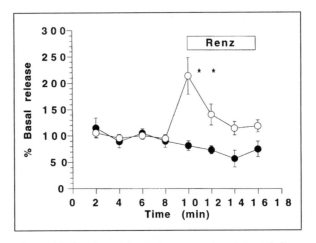

Fig. 5.3. Modulation of dopamine release in rat striatal slices via the 5-HT$_4$ receptor. The 5-HT$_4$ receptor agonist renzapride (O; 10 μM; Renz) increased the release of dopamine in rat striatal slices and this response was prevented by prior administration of the selective 5-HT$_4$ receptor antagonist GR113808 (●; 100 nM present throughout the collection period). GR113808 (100 nM) administered alone failed to modify basal dopamine release.[55] Dopamine levels are expressed as the percentage of the mean absolute amount in the 4 collections proceeding the drug treatment. Data represent mean ± SEM, n = 3 - 14. Horizontal bar represents application of renzapride. ANOVA<0.05, **P<0.01 (Dunnett's t test). Reproduced with permission from Steward LJ et al, Br J Pharmacol 1996; 117:55-62.

release of striatal dopamine in anesthetized[53,54] and freely moving rats in vivo[55] (Fig. 5.2), and also using slices of rat striatum[55] (Fig. 5.3).

The selective 5-HT$_4$ receptor antagonist, GR113808, administered alone tended to reduce the release of striatal dopamine in freely moving rats, assessed by the microdialysis technique, suggesting the presence of an endogenous tone on the 5-HT$_4$ receptor in this preparation[55] (Fig. 5.2). This action of GR113808, however, was not apparent in striatal slices which presumably reflects lower levels of endogenous tone due to a reduction in 5-HT neurone terminal activity relative to the in vivo preparation[55] (Fig. 5.3). Furthermore, neither the nonselective 5-HT$_{3/4}$ receptor antagonist, DAU6285, nor the selective 5-HT$_4$ receptor antagonist, GR125487, altered basal striatal dopamine release in anesthetized rats.[52,56] Again, a reduction in the activity of neurones, in this latter case induced by the anesthetic agent, would decrease the endogenous tone on the $5\text{-}HT_4$ receptor in this preparation.

The $5\text{-}HT_4$ receptor-mediated increase in striatal dopamine release in vitro and in vivo was prevented by application of the membrane permeable phospholipase C/cAMP-dependent protein kinase A inhibitor H7,[57] indicating that the response was mediated via a metabotropic-like receptor,[55] consistent with the known transduction system associated with the $5\text{-}HT_4$ receptor. In addition, the reduction in the ability of 5-HT to enhance striatal dopamine release in the presence of forskolin (which alone enhanced dopamine release[58]), may be due to the already enhanced activity of adenylate cyclase induced by this direct adenylate cyclase activator.

The $5\text{-}HT_4$ receptor agonist-induced increase in striatal dopamine release in vitro and in vivo, is prevented by the inclusion of the Na^+ channel blocker, tetrodotoxin[54,55] (Fig. 5.4). Furthermore, the $5\text{-}HT_4$ receptor-mediated response is not apparent in striatal synaptosomes preloaded with either [³H]tyrosine (to allow a measure of endogenously synthesized [³H]dopamine) or [³H]dopamine.[56] The corollary of these studies is that the $5\text{-}HT_4$ receptor-mediated modulation of striatal dopamine release is indirect (i.e., the $5\text{-}HT_4$ receptor is not located on dopamine neurone terminals in the striatum). It is therefore of interest that autoradiographic studies have demonstrated that $5\text{-}HT_4$ receptor levels are not altered in the striatum of rat brain following 6-hydroxydopamine lesion of the nigral-striatal dopamine system.[39,40] This indicates that at least a major population of $5\text{-}HT_4$ receptors in the striatum is not located on dopamine neurone terminals consistent with the functional studies. Interestingly, comparable findings have been demonstrated with human brain tissue from patients with Parkinson's disease which is neuropathologically characterized by a lesion of the dopaminergic nigro-striatal pathway.[59]

The phenotype(s) of the neurones which express $5\text{-}HT_4$ receptors in the striatum remains to be determined, however, the substantial reduction (80-90% decrease) in radiolabelled striatal $5\text{-}HT_4$ receptors following intrastriatal administration of the neurotoxin, kainic acid,[39,40] indicates that at least a major population of $5\text{-}HT_4$ receptors in the striatum are expressed by neurones that have their cell bodies in this region. Consistent findings have been demonstrated with human brain tissue by using patients with Huntington's disease[59]

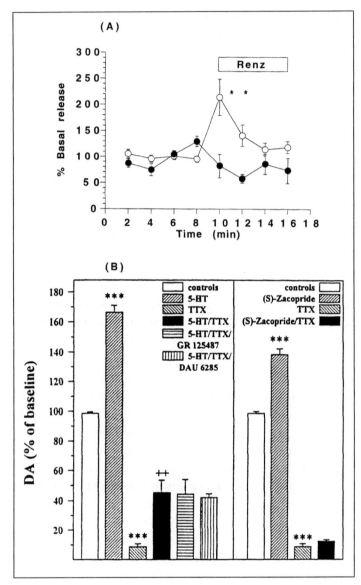

Fig. 5.4. Ability of the Na$^+$ channel blocker tetrodotoxin to prevent the 5-HT$_4$ receptor mediated increase in striatal dopamine release in vitro and in vivo. A: Continuous perfusion of tetrodotoxin prevented the increase in dopamine release from rat striatal slices induced by the 5-HT$_4$ receptor agonist renzapride (Renz; 10 μM; application indicated by bar). Renzapride alone (O), renzapride plus tetrodotoxin (●). Tetrodotoxin (100 nM) administered alone failed to modify basal dopamine release.[55] Dopamine levels are expressed as the percentage of the meaned absolute amount in the 4 collections proceeding the drug treatment. Data represent mean ± S.E. mean, n = 5 - 14. ANOVA<0.05, **P<0.01 (Dunnett's t test). B: Continuous perfusion of tetrodotoxin (1 μM) reduced and abolished the ability of 5-HT and (S)-zacopride (1 and 100 μM, administered via the microdialysis probe for 15 min, respectively) to increase dopamine release in the

which is neuropathologically characterized by a major loss of striatal neurones.[60] Indeed, the small reduction (20-40% decrease) in 5-HT$_4$ receptor levels in the ipsilateral substantia nigra following the intrastriatal kainic acid lesion indicates that some of the 5-HT$_4$ receptors in the substantia nigra are on the GABAergic terminals originating from the striatum.[39,40] However, the presence of 5-HT$_4$ receptor mRNA within the substantia nigra pars reticulata[17] indicates that 5-HT$_4$ receptors may also be on additional neurones within this region. Furthermore, the reduction (~70% decrease) in 5-HT$_4$ receptor levels within the ipsilateral globus pallidus following intrastriatal kainic acid lesion indicates that 5-HT$_4$ receptors are also located on the GABAergic terminals in this region which originate in the striatum.[40] This latter finding, however, was not evident in an earlier study.[39] It would appear, however, that a major population of 5-HT$_4$ receptors are not expressed by either central 5-HT neurones or the striatal terminals of cortical-striatal glutamate projection neurones since neither central 5-HT neurone destruction nor lesion of the glutamatergic cortical-striatal projection reduced levels of striatal 5-HT$_4$ receptors.[40]

Given that 5-HT$_4$ receptor activation results in an indirect increase in striatal dopamine release and that activation of neuronally expressed 5-HT$_4$ receptors is likely to yield neuronal excitation, at least two potential mechanisms may be forwarded which are consistent with the lesion studies (Fig. 5.5). Thus, 5-HT$_4$ receptors located on the striatal elements of the striatal-pallidal GABAergic projection would facilitate the inhibition of the GABAergic pallidal-nigral projection. This would lead to an increase in the disinhibition of the nigral-striatal dopamine projection, resulting in an increase in striatal dopamine release (see refs. 61-63). However, such a mechanism

striatum of anesthetized rats assessed using the in vivo microdialysis technique. The failure of the selective and nonselective 5-HT$_4$ receptor antagonists GR125487 and DAU6285 (1 µM and 100 µM, administered via the microdialysis probe 45 min prior to and during 5-HT application, respectively) to attenuate the action of 5-HT in the presence of tetrodotoxin indicates that this response is not mediated via the 5-HT$_4$ receptor but more likely reflects competition by 5-HT for the dopamine uptake channel (see ref. 52). Each column represents the mean ± SEM, n = 4-5 at the time of maximum agonist-induced effect. [++]P<0.01, two-way ANOVA, [***]P<0.001, versus control (Tukey's test). GR125,487 (1 µM) and DAU6285 (100 µM) did not significantly modify the tetrodotoxin-insensitive component of the 5-HT-induced response (one-way ANOVA). Reproduced with permission from Steward LJ et al, Br J Pharmacol 1996; 117:55-62 and DeDeurwaerdère P et al, J Neurochem 1997; 68:195-203.

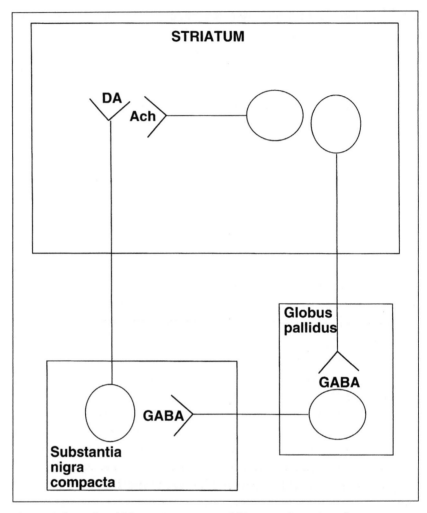

Fig. 5.5. Pathways by which 5-HT$_4$ receptor could increase dopamine release.

would only be relevant to explain the in vivo studies since in a striatal slice preparation, these pathways would not be intact. As an alternative, 5-HT$_4$ receptor stimulation of striatal cholinergic neurones would cause a local release of acetylcholine which could then activate either striatal muscarinic or nicotinic acetylcholine receptors; activation of either of these receptor subtypes has previously been shown to enhance striatal dopamine release (e.g., ref. 64, 65). As mentioned previously, however, Consolo and colleagues[31] failed to detect an increase in striatal acetylcholine release following 5-HT$_4$

receptor activation. However, in these studies drugs were administered i.c.v. rather than intrastriatal, and also the acetylcholine levels were enhanced by the inclusion of the acetylcholinesterase inhibitor, physostigmine. These modifications may have prevented detection of the response.

In a further study investigating the ability of the 5-HT$_4$ receptor to modulate forebrain dopamine release, the selective 5-HT$_4$ receptor antagonist, GR113808 (1-10 µM), administered via a microdialysis probe, failed to modify dopamine release in either the rat striatum or nucleus accumbens, although administration of a higher concentration (100 µM) reduced dopamine release in the striatum (but not nucleus accumbens).[66] As indicated by the authors, this latter response was unlikely to be mediated via the 5-HT$_4$ receptor since such high concentrations of GR113808 were required to evoke the response (GR113808 K$_i$ for 5-HT$_4$ receptors $\leq$ 1 nM),[38] even when a recovery factor for the drug to cross the dialysis membrane is taken into account (typically 10-20% of the drug will cross the dialysis membrane and enter the brain). In the same study, when dopamine release in the striatum and nucleus accumbens was facilitated by systemic administration of morphine, a response due to the disinhibition of dopamine neurones mediated by inhibitory µ-opiate receptors on GABAergic neurones (for review see ref. 67), subsequent bilateral intranigral injection of GR113808 (10 µg but not 1 or 2.5 µg) attenuated the elevation in dopamine release in the striatum.[67] While the authors proposed that this latter GR113808-induced response was 5-HT$_4$ receptor-mediated, given that the dose of GR113808 would give a concentration of over 100 µM in 100 µl (which is a considerably larger volume than the substantia nigra), and that only 4- to 10-fold lower doses of GR113808 were ineffective, further investigation is warranted prior to ascribing a modulatory role for 5-HT$_4$ receptors in this response. Indeed, at such high doses, GR113808 would be expected to interact with 5-HT$_3$ receptors (GR113808 K$_i$ for 5-HT$_3$ receptors = 1 µM),[36-38] and 5-HT$_3$ receptor antagonists are known to attenuate the morphine-induced dopamine increase in the nucleus accumbens upon both peripheral and mesencephalic administration.[68-70]

In contrast to the numerous reports indicating that the 5-HT$_4$ receptor modulates striatal dopamine release, administration of 5-HT$_4$ receptor ligands produces no overt behavioral change (e.g., ref. 46). Furthermore, administration of a brain penetrant 5-HT$_4$

receptor antagonist failed to modulate a number of behaviors which are likely to result from enhanced dopamine release.[71,72] This may be explained by the low level of endogenous tone on the central $5\text{-}HT_4$ receptor, in concordance with which, central $5\text{-}HT_4$ receptor antagonist administration at most induces only a modest reduction in dopamine release (Fig. 5.2).[53,55] However, the failure of the lipophilic $5\text{-}HT_4$ receptor partial agonist, RS67333, at centrally active doses, to modify locomotor activity (estimated by swim speed[46]) further complicates the issue. It should be noted, however, that while the $5\text{-}HT_4$ receptor antagonist, RS67532, did not modify swim speed, this compound reduced locomotor activity measured in activity boxes at the same dose as used to antagonize the $5\text{-}HT_4$ receptor-mediated facilitation of cognitive performance.[46] Since central dopaminergic neurotransmission is clearly associated with locomotor activity (e.g., ref. 73), more work in this area is needed to further clarify the role of the $5\text{-}HT_4$ receptor in the behavioral expression of locomotor activity.

$5\text{-}HT_4$ Receptor-Mediated Modulation of Central GABA Release

As previously mentioned, lesion studies indicate that striatal-pallidal and striatal-nigral projection neurones (presumably GABAergic neurones[62]) express $5\text{-}HT_4$ receptors. Consistent with this evidence, the release of GABA in the substantia nigra of anesthetized rats has been shown to be modulated by the $5\text{-}HT_4$ receptor.[74] However, only $5\text{-}HT_4$ receptor antagonists (GR113808 and SB204070) evoke a response and this was only evident under depolarizing conditions (elevated $[K^+]$). Thus administration of either antagonist failed to modify basal GABA release in the substantia nigra, but when neuronal activity was elevated by the perfusion of aCSF containing a higher concentration of KCl (30 mM), the increase in GABA release was attenuated by both $5\text{-}HT_4$ receptor antagonists.[74] These results suggest that, under depolarizing conditions, the $5\text{-}HT_4$ receptors are tonically active and result in an enhanced release of GABA.[74]

$5\text{-}HT_4$ Receptor-Mediated Modulation of Central 5-HT Release

It has been known for some time that, unlike many selective $5\text{-}HT_3$ receptor antagonists, the $5\text{-}HT_4$ receptor agonist/$5\text{-}HT_3$ receptor antagonists (e.g., renzapride, S(-)-zacopride) fail to display anxiolytic-like action in animal models, suggesting that the $5\text{-}HT_4$

receptor agonism prevents the behavioral action due to selective 5-HT$_3$ receptor antagonism (for review see ref. 75). Furthermore, unlike selective 5-HT$_3$ receptor antagonists, compounds such as renzapride and (S)-zacopride enhance central 5-HT release,[76-77] which would be expected to enhance levels of 'anxiety' (e.g., refs. 78-82) or possibly, as is consistent with the behavioral data, prevent an anxiolytic-like response due to antagonism of the 5-HT$_3$ receptor.[75] However, the ability of S(-)zacopride to reverse the anxiolytic-like response of the 5-HT synthesis inhibitor PCPA (parachlorophenylalanine),[83,84] an action not mimicked by the selective 5-HT$_3$ receptor antagonist ondansetron,[83] suggests that this 5-HT$_4$ receptor agonist may (also?) act distally to the 5-HT neurone. Although it should be noted that in these latter studies, only a moderate dose of PCPA was employed that reduced central 5-HT levels by less than 20%,[84] which may not be low enough to eliminate responses mediated via 5-HT neurones.

Given the known influence of 5-HT in the hippocampus on the expression of anxiety (e.g., refs. 78-81), we subsequently assessed the effects of intrahippocampal administration of the 5-HT$_4$ receptor agonist (and 5-HT$_3$ receptor antagonist), renzapride, on monoamine neurotransmitter release in this brain region. Renzapride facilitated the release of 5-HT in the hippocampus of freely moving rats assessed using the microdialysis technique[85] (Fig. 5.6); an effect that was mimicked by the nonselective 5-HT$_4$ receptor agonist, 5-methoxytryptamine (in presence of pindolol and methysergide to antagonize some of the 5-methoxytryptamine-sensitive non-5-HT$_4$ 5-HT receptors).[85] Furthermore, the 5-HT$_4$ receptor agonist-facilitated release was prevented by prior intrahippocampal administration of either of the selective 5-HT$_4$ receptor antagonists, GR113808 or GR125487 (Fig. 5.6), supporting the involvement of the 5-HT$_4$ receptor in the response. The latter 5-HT$_4$ receptor antagonist was also effective when administered peripherally,[85] indicating that it enters the brain following systemic administration (although the implantation of a microdialysis probe may disrupt the blood brain barrier), which is consistent with results reported recently.[47]

It is also noteworthy that both 5-HT$_4$ receptor antagonists (GR113808 and GR125487) were able to considerably reduce the release of 5-HT in the hippocampus, indicating the presence of an endogenous tone on the 5-HT$_4$ receptor within this preparation (Fig. 5.6).[85] The magnitude of the 5-HT$_4$ receptor antagonist-induced

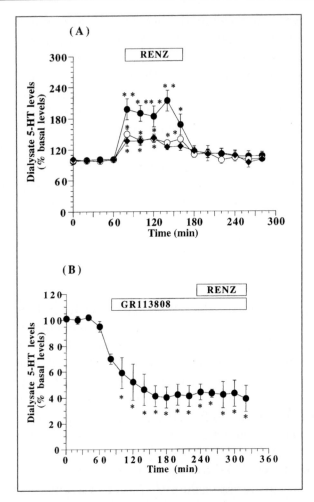

Fig. 5.6. Modulation of 5-HT release in rat hippocampus via the 5-HT₄ receptor, in vivo. The 5-HT₄ receptor agonist renzapride increased 5-HT release in the rat hippocampus assessed using the in vivo microdialysis technique (A; renzapride, RENZ: ◆; 1.0 μM, ○; 10 μM, ●; 100 μM, administered via the perfusing aCSF) and (B) prevention of the response by prior and continued application of the selective 5-HT₄ receptor antagonist GR113808 (100 nM). The horizontal bars represent application of the indicated drugs. Data represent the mean ± S.E. mean, n=4-6. ANOVA P<0.05, *P<0.05, **P<0.01 (Dunnett's t test). Reproduced with permission from Ge J et al, Br J Pharmacol 1996; 117:1475-1480.

reduction in 5-HT release was surprisingly large (being comparable to that induced by a near maximal dose of the somatodendritic 5-HT$_{1A}$ autoreceptor agonist 8-OHDPAT.[86] However, since our studies were performed on awake animals under standard laboratory lighting, with the animals being placed in a novel environment in the vicinity of an investigator to collect the dialysate samples, it is likely that these animals were sufficiently 'anxious' to considerably increase the release of 5-HT in the hippocampus (e.g., refs. 78-81), which would subsequently enhance the tone on the 5-HT$_4$ receptor.

Our findings that 5-HT$_4$ receptor antagonists decrease 5-HT release concur with recently reported studies indicating that 5-HT$_4$ receptor antagonists possess anxiolytic-like effects,[72,86] which is consistent with the association of central 5-HT function with anxiety, i.e., a reduction in central 5-HT function reduces levels of anxiety (e.g., refs. 87, 88). While this theory may not be relevant for all the 5-HT receptors, nor indeed for all pathologies that manifest the symptoms of anxiety, the ability of 5-HT$_4$ receptor agonists to increase (and 5-HT$_4$ receptor antagonists to decrease) 5-HT release in the dorsal hippocampus,[85] provides a relevant neurochemical mechanism to explain the 5-HT$_4$ receptor-mediated modulation of 'anxiety'. It should be noted, however, that another group have consistently demonstrated an opposite behavioral effect of 5-HT$_4$ receptor antagonists.[89-91] Clearly, further studies utilizing a range of 5-HT$_4$ receptor agonists and antagonists will help clarify the situation concerning the apparent inconsistencies relating to the involvement of the 5-HT$_4$ receptor with the expression of anxiety.

Despite the ability of the 5-HT$_4$ receptor to modulate 5-HT release in the hippocampus, no other evidence is available to indicate that the 5-HT$_4$ receptor is expressed by 5-hydroxytryptaminergic neurone terminals in this brain region. Thus, 5,7-dihydroxytryptamine lesion of 5-HT neurones resulted in an upregulation of 5-HT$_4$ receptor levels in many forebrain structures, including the hippocampus, indicating that 5-HT$_4$ receptors are primarily expressed postsynaptically to the 5-HT neurone.[40] Furthermore, no studies have reported 5-HT$_4$ receptor mRNA in the raphé nuclei which contain the 5-HT neurone cell bodies that project to the forebrain. However, the apparently close association of 5-HT$_4$ receptor mRNA and 5-HT$_4$ receptor binding sites with cell layers that predominantly possess glutamatergic neurones[17,40,43,44,92,93] and functional data indicating that 5-HT$_4$ receptors excite hippocampal pyramidal neurones[24] suggests that $5\text{-}HT_4$ receptor activation results in the neuronal release of glutamate in the hippocampus which may subsequently excite 5-HT neurone terminals.

Conclusion

The availability of a number of selective $5\text{-}HT_4$ receptor ligands has considerably benefited research investigating the role of the $5\text{-}HT_4$ receptor. Hence the utilization of these pharmacological tools has clearly demonstrated that the brain expresses functional $5\text{-}HT_4$ receptors. Furthermore, while this area is still in its infancy, there is increasing optimism that central $5\text{-}HT_4$ receptors may provide therapeutic targets for the alleviation of a number of CNS disorders (e.g., anxiety, memory impairment). It remains to be determined whether this optimism is justified.

Acknowledgments

We are grateful to Drs. A. J. Fulford and R. L. Stowe for comments on the manuscript. Research in NMB's laboratory is supported by the MRC, the Wellcome Trust and the British Pharmacological Society.

References

1. Boess FG, Martin IL. Molecular biology of 5-HT receptors. Neuropharmacology 1994; 33:275-317.
2. Hoyer D, Clarke DE, Fozard JR, Hartig PR, Martin GR, Mylecharane EJ, Saxena PR, Humphrey PPA. International Union of Pharmacology classification of receptors for 5-hydroxytryptamine (serotonin). Pharmacol Rev 1994; 46:157-203.
3. Von Hungren K, Roberts S, Hill DF. Developmental and regional variations in neurotransmitter-sensitive adenylate cyclase systems in cell-free preparations from rat brain. J Neurochem 1974; 22:811-819.
4. Von Hungren K, Roberts S, Hill DF. Serotonin-sensitive adenylate cyclase activity in immature rat brain. Brain Res 1975; 84:257-267.
5. Enjalbert A, Bourgoin S, Hamon M, Adrien J, Bockaert J. Postsynaptic serotonin-sensitive adenylate cyclases in the central nervous system I. Development and distribution of serotonin and dopamine-sensitive adenylate cyclases in rat and guinea pig brain. Mol Pharmacol 1978a; 14:2-10.
6. Enjalbert A, Hamon M, Bourgoin S, Bockaert J. Postsynaptic serotonin-sensitive adenylate cyclases in the central nervous system II. Comparison with dopamine- and isoproterenol-sensitive adenylate cyclases in rat brain. Mol Pharmacol 1978b; 14:11-23.
7. Dumuis A, Bouhelal R, Sebben M, Bockaert J. A 5-HT receptor in the central nervous system positively coupled with adenylate cyclase is antagonised by ICS 205-930. Eur J Pharmacol 1988a; 146:187-188.

8. Dumuis A, Bouhelal R, Sebben M, Cory R, Bockaert J. A nonclassical 5-hydroxytryptamine receptor positively coupled with adenylate cyclase in the central nervous system. Mol Pharmacol 1988b; 34:880-887.

9. Dumuis A, Sebben M, Bockaert J. The gastrointestinal prokinetic benzamide derivatives are agonists at the non-classical 5-HT receptor (5-HT$_4$) positively coupled to adenylate cyclase in neurons. Naunyn-Schmiedeberg's Arch Pharmacol 1989; 340:403-410.

10. Bockaert J, Sebben M, Dumuis A. Pharmacological characterisation of 5-hydroxytryptamine$_4$ (5-HT$_4$) receptors positively coupled to adenylate cyclase in adult guinea pig hippocampal membranes: effect of substituted benzamide derivatives. Mol Pharmacol 1990; 37:408-411.

11. Shenker A, Maayani S, Weinstein H, Green JP. Pharmacological characterisation of two 5-hydroxytryptamine receptors coupled to adenylate cyclase in guinea pig hippocampal membranes. Mol Pharmacol 1987; 31:357-367.

12. Ford APDW, Clarke DE. The 5-HT$_4$ Receptor. Med Res Rev 1993; 13:633-662.

13. Monferini E, Gaetani P, Rodriguez y Baena R, Giraldo E, Parenti M, Zocchetti R, Rizzi CA. Pharmacological characterisation of the 5-hydroxytryptamine receptor coupled to adenylyl cyclase stimulation in human brain. Life Sci 1993; 52:61-65.

14. Gerald C, Hartig PR, Branchek TA, Weinshank RL. DNA encoding 5-HT$_4$ serotonin receptors and uses thereof. International Patent Number WO 94/14957 1994.

15. Gerald C, Adham A, Kao HT, Olsen MA, Laz TM, Schechter LE, Bard JA, Vaysse PJJ, Hartig PR, Branchek TA, Weinshank RL. The 5-HT$_4$ receptor: molecular cloning and pharmacological characterisation of two splice variants. EMBO J 1995; 14:2806-2815.

16. Claeysen S, Sebben M, Journot L, Bockaert J, Dumuis A. Cloning, expression and pharmacology of the mouse 5-HT$_{4L}$ receptor. FEBS Lett 1996; 398:19-25.

17. Vilaró MT, Cortés R, Gerald C, Branchek TA, Palacios JM, Mengod G. Localization of 5-HT$_4$ receptor mRNA in rat brain by in situ hybridisation histochemistry. Mol Brain Res 1996; 43:356-360.

18. Ansanay H, Sebben M, Bockaert J, Dumuis A. Characterisation of homologous 5-hydroxytryptamine$_4$ receptor desensitisation in colliculi neurons. Mol Pharmacol 1992; 42:808-816.

19. Ansanay H, Sebben M, Bockaert J, Dumuis A. Pharmacological comparison between [^{3}H]GR113808 binding sites and functional 5-HT$_4$ receptors in neurons. Eur J Pharmacol 1996; 298:165-174.

20. Rondé P, Ansanay H, Dumuis A, Miller R, Bockaert J. Homologous desensitization of 5-hydroxytryptamine$_4$ receptors in rat esophagus: functional and second messenger studies. J Pharmacol Exp Ther 1995; 272:977-983.

21. Fagni L, Dumuis A, Sebben M, Bockaert J. The 5-HT$_4$ receptor subtype inhibits K$^+$ current in colliculi neurones via activation of a cy-

clic AMP-dependent protein kinase. Br J Pharmacol 1992; 105: 973-979.

22. Chaput Y, Aranda RC, Andrade R. Pharmacological and functional analysis of a novel serotonin receptor in the rat hippocampus. Eur J Pharmacol 1990; 182:441-456.

23. Andrade R, Chaput Y. 5-Hydroxytryptamine₄-like receptors mediate the slow excitatory response to serotonin in the rat hippocampus. J Pharmacol Exp Ther 1991; 257:930-937.

24. Roychowdhury S, Haas H, Anderson EG. 5-HT₁ₐ and 5-HT₄ receptor colocalization on hippocampal pyramidal cells. Neuropharmacology 1994; 33:551-557.

25. Tonini M, Galligan JJ, North RA. Effects of cisapride on cholinergic neurotransmission and propulsive motility in the guinea-pig ileum. Gastroenterology 1989; 96:1257-1264.

26. Tonini M, Stefano CM, Onori L, Coccini T, Manzo L, Rizzi CA. 5-Hydroxytryptamine₄ receptor agonists facilitate cholinergic transmission in the circular muscle of guinea-pig ileum: antagonism by tropisetron and DAU6285. Life Sci 1992; 50:173-178.

27. Craig DA, Clarke DE. Pharmacological characterisation of a neuronal receptor for 5-hydroxytryptamine in guinea-pig ileum with properties similar to the 5-hydroxytryptamine₄ receptor. J Pharmacol Exp Ther 1990; 252:1378-1386.

28. Eglen RM, Swank SR, Walsh LKM, Whiting RL. Characterisation of 5-HT₃ and 'atypical' 5-HT receptors mediating guinea-pig ileal contractions in vitro. Br J Pharmacol 1990; 101:513-520.

29. Elswood CJ, Bunce KT, Humphrey PPA. Identification of putative 5-HT₄ receptors in guinea-pig ascending colon. Eur J Pharmacol 1991; 196:149-155.

30. Kilbinger H, Gebauer A, Haas J, Ladinsky H, Rizzi CA. Benzimidazolones and renzapride facilitate acetylcholine release from guinea-pig myenteric plexus via 5-HT₄ receptors. Naunyn-Schmiedebergs Arch. Pharmacol 1995; 351:229-236.

31. Consolo S, Arnaboldi S, Giorgi S, Russi G, Ladinsky H. 5-HT(4) receptor stimulation facilitates acetylcholine release in rat frontal cortex. Neuroreport 1994; 5:1230-1232.

32. Baxter GS, Clarke DE. Benzimidazolone derivatives act as 5-HT₄ receptor ligands in rat oesophagus. Eur J Pharmacol 1992; 212:225-229.

33. Dumuis A, Sebben M, Monferini E, Nicola M, Turconi M, Ladinsky H, Bockaert J. Azabicycloalkyl benzimidazolone derivatives as a novel class of potent agonists at the 5-HT₄ receptor positively coupled to adenylate cyclase in brain. Naunyn-Schmiedeberg's Arch Pharmacol 1991; 343:245-251.

34. Medhurst AD, Kaumann AJ. Characterisation of the 5-HT₄ receptor mediating tachycardia in piglet isolated right atrium. Br J Pharmacol 1993; 110:1023-1030.

35. Bockaert J, Fozard JR, Dumuis A, Clarke DE. The 5-HT₄ receptor: a place in the sun. Trends Pharmacol Sci 1992; 13:141-145.

36. Gale JD, Green A, Darton J, Sargent RS, Clayton NM, Bunce KT. GR125487: A 5-HT$_4$ receptor antagonist with a long duration of action in vivo. Br J Pharmacol 1994a; 113:119P.

37. Gale JD, Grossman CJ, Darton J, Bunce KT, Whitehead JWF, Knight J, Parkhouse TJ, Oxford AW, Humphrey PPA. GR125487: A selective and high affinity 5-HT$_4$ receptor antagonist. Br J Pharmacol 1994b; 113:120P.

38. Gale JD, Grossman CJ, Whitehead JWF, Oxford AW, Bunce KT, Humphrey PPA. GR113808—A Novel, Selective Antagonist with High Affinity at the 5-HT$_4$ Receptor. Br J Pharmacol 1994c; 111:332-338.

39. Patel S, Roberts J, Moorman J, Reavil C. Localization of serotonin-4 receptors in the striatonigral pathway in rat brain. Neuroscience 1995; 69:1159-1167.

40. Compan V, Daszuta A, Salin P, Sebben M, Bockaert J, Dumuis A. Lesion study of the distribution of serotonin 5-HT$_4$ receptors in rat basal ganglia and hippocampus. Eur J Neurosci 1996; 8:2591-2598.

41. Boddeke HW, Kalkman HO. Zacopride and BRL24924 induce an increase in EEG-energy in rats. Br J Pharmacol 1990; 101:281-284.

42. Boddeke HW, Kalkman HO. Agonist effects at putative central 5-HT$_4$ receptors in rat hippocampus by R(+)- and S(-)-zacopride: no evidence for stereoselectivity. Neurosci Lett 1992; 134:261-263.

43. Ullmer C, Engels P, Abdel'Al S, Lübbert H. Distribution of 5-HT$_4$ receptor mRNA in the rat brain. Naunyn-Schmiedeberg's Arch Pharmacol 1996; 354:210-212.

44. Mengod G, Vilaró MT, Raurich A, López-Giménez JF, Cortés R, Palacios JM. 5-HT receptors in mammalian brain: receptor autoradiography and in situ hybridisation studies of new ligands and newly identified receptors. Histochem J 1996; 28:747-758.

45. Bartus RT, Dean RL, Beer B, Lippa AS. The cholinergic hypothesis of geriatric memory dysfunction. Science 1982; 217:408-417.

46. Fontana DJ, Daniels SE, Wong EHK, Clark RD, Eglen RM. The effects of novel, selective 5-hydroxytryptamine (5-HT)$_4$ receptor ligands in rat spatial navigation. Neuropharmacology 1997; 36:689-696.

47. Letty S, Child R, Dumuis A, Pantaloni A, Bockaert J, Rondouin G. 5-HT$_4$ receptors improve social olfactory memory in rat. Neuropharmacology 1997; 36:681-687.

48. Marchetti-Gauthier E, Roman FS, Dumuis A, Bockaert J, Soumireu-Mourat B. BIMU1 increases associative memory in rats by activating 5-HT$_4$ receptors. Neuropharmacology 1997; 36:697-706.

49. Meneses A, Hong E. Effect of 5-HT$_4$ receptor agonists and antagonists in learning. Pharmacol Biochem Behav 1997; 56:347-351.

50. Bliss TVP, Collingridge GL. A synaptic model of memory—long-term potentiation in the hippocampus. Nature 1993; 361:31-39.

51. Benloucif S, Keegan MJ, Galloway MP. Serotonin-Facilitated Dopamine Release In vivo—Pharmacological Characterisation. J Pharmacol Exp Ther 1993; 265:373-377.

52. DeDeurwaerdère P, Bonhomme N, Lucas G, Le Moal M, Spampinato U. Serotonin enhamces striatal dopamine outflow in vivo through dopamine uptake sites. J Neurochem 1996; 66:210-215.

53. Bonhomme N, De Deurwaerdere P, Le Moal M, Spampinato U. Evidence for 5-HT₄ receptor subtype involvement in the enhancement of striatal dopamine release induced by serotonin: a microdialysis study in the halothane-anesthetized rat. Neuropharmacology 1995; 34:269-279.

54. DeDeurwaerdéré P, L'Hirondel M, Bonhomme N, Lucas G et al. Serotonin stimulation of 5-HT₄ receptors indirectly enhances in vivo dopamine release in the rat striatum. J Neurochem 1997; 68:195-203.

55. Steward LJ, Ge J, Stowe RL, Brown DC, Bruton RK, Stokes PRA, Barnes NM. Ability of 5-HT₄ receptor ligands to modulate rat striatal dopamine release in vitro and in vivo. Br J Pharmacol 1996; 117:55-62.

56. DeDeurwaerdère P, L'hirondel M, Bonhomme N, Lucas G, Cheramy A, Spampinato U. Serotonin stimulation of 5-HT₄ receptors indirectly enhances in vivo dopamine release in the rat striatum. J Neurochem 1997; 68:195-203.

57. Hidaka H, Inayaki M, Kawamoto S, Sasaki Y. Isoquinolinesulfonamides, novel and potent inhibitors of cyclic nucleotide dependent protein kinase and protein kinase C. Biochemistry 1984; 23:5036-5046.

58. West AR, Galloway MP. Desensitization of 5-hydroxytryptamine-facilitated dopamine release in vivo. Eur J Pharmacol 1996; 298:241-245.

59. Reynolds GP, Mason SL, Meldrum A, Dekeczer S, Parnes H, Eglen RM, Wong EHF. 5-Hydroxytryptamine (5-HT)₄ receptors in post mortem human brain tissue: Distribution, pharmacology and effects of neurodegenerative diseases. Br J Pharmacol 1995; 114:993-998.

60. Bird ED. Huntington's disease In: Riederer P, Kopp N, Pearson J, eds. An Introduction to Neurotransmission in Health and Disease. Oxford Medical Publications, 1990:221-233.

61. Gerfen CR. The neostriatal mosaic: multiple levels of compartmental organization. Trends Neurosci 1992; 15:133-139.

62. Kawaguchi Y, Wilson CJ, Augood SJ, Emson PC. Striatal interneurones: chemical, physiological and morphological characterization. Trends Neurosci 1995; 18:527-535.

63. Bevan MD, Smith AD, Bolam JP. The substantia nigra as a synaptic integration of functionally diverse information arising from the ventral pallidum and the globus pallidus in the rat. Neuroscience 1996; 75:5-12.

64. Besson MJ, Kemel ML, Glowinski J, Giorguieff MF. Presynaptic control of dopamine release from striatal dopaminergic terminals by various striatal transmitters. In: Langer SZ, Starke K, Dubocovich ML, eds. Presynaptic Receptors, Advances in the Biosciences (volume 18). New York: Pergamon Press, 1978:159-163.

65. DeBelleroche J, Bradford HF. Presynaptic control of the synthesis and release of dopamine from striatal synaptosomes: a comparison between the effects of 5-hydroxytryptamine, acetylcholine and glutamate. J Neurochem 1980; 35:1227-1234.

66. Pozzi L, Trabace L, Invernizzi R, Samanin R. Intranigral GR-113808, a selective 5-HT₄ receptor antagonist, attenuates morphine-stimu-

lated dopamine release in the rat striatum. Brain Res 1995; 692:265-268.

67. DiChiara G, North RA. Neurobiology of opiate abuse. Trends Pharmacol Sci 1992; 13:185-193.

68. Carboni E, Acquas E, Frau R, DiChiara, G. Differential inhibitory effects of a 5-HT$_3$ antagonist on drug-induced stimulation of dopamine release. Eur J Pharmacol 1989; 164:515-519.

69. Imperato A, Angelucci L. (1989) 5-HT$_3$ receptors control dopamine release in the nucleus accumbens of freely moving rats. Neurosci Lett 101:214-217

70. Pei Q, Zetterström T, Leslie RA, Grahame-Smith DG. 5-HT$_3$ receptor antagonists inhibit morphine-induced stimulation of mesolimbic dopamine release and function in the rat. Eur J Pharmacol 1993; 230:63-68.

71. Reavil C, Hatcher JP, Zetterstrom TSC, Lewis AV, Sanger GJ, Hagan JJ, McKay S. 5-HT$_4$ receptor antagonism does not affect dopamine-mediated behavioral effects in the rat. Br J Pharmacol 1996; 118(suppl):71P.

72. Kennett GA, Bright F, Trail B, Blackburn TP, Sanger GJ. Anxiolytic-like actions of the selective 5-HT$_4$ receptor antagonists, SB204070A and SB207266A in rats. Neuropharmacology 1997; 36:707-712.

73. Eison AS, Eison, MS, Iversen, SD. The behavioral effects of a novel substance P analogue following infusion into the ventral tegmental area of the substantia nigra of the rat brain. Brain Res 1982; 238:137-152.

74. Zetterström TSC, Husum H, Smith S, Sharp T. Local application of 5-HT$_4$ antagonists inhibits potassium-stimulated GABA efflux from rat substantia nigra in vivo. Br J Pharmacol 1996; 119(suppl):347P.

75. Bentley KR, Barnes NM. Therapeutic potential of 5-HT$_3$ receptor antagonists in neuropsychiatric disorders. CNS Drugs 1995; 3:363-392.

76. Barnes NM, Cheng CHK, Costall B, Ge J, Naylor RJ. Differential modulation of extracellular levels of 5-hydroxytryptamine in the rat frontal cortex by (R)- and (S)-zacopride. Br J Pharmacol 1992b; 107:233-239.

77. Ge J, Barnes NM, Cheng CHK, Costall B, Naylor RJ. Interaction of (R)- and (S)-zacopride and 5-HT$_3$/5-HT$_4$ receptor ligands to modify extracellular levels of 5-HT in the rat frontal cortex. Br J Pharmacol 1992; 107(suppl):111P, 1992.

78. Andrews N, File SE. Increased 5-HT release mediates the anxiogenic response during benzodiazepine withdrawal—A review of supporting neurochemical and behavioral evidence. Psychopharmacology 1993; 112:21-25.

79. Andrews N, Hogg S, Gonzalez LE, File S. 5-HT$_{1A}$ receptors in the median raphe and dorsal hippocampus may mediate anxiolyic and anxiogenic behaviors respectively. Eur J Pharmacol 1994; 264:259-264.

80. Andrews N, File SE, Fernandes C, Gonzalez LE, Barnes NM. Evidence that the median raphe nucleus—dorsal hippocampal pathway

mediates diazepam withdrawal-induced anxiety. Psychopharmacology 1997; 130:228-234.

81. Cadogan AK, Kendall DA, Fink H, Marsden CA. Social interaction increases 5-HT release and cAMP efflux in the rat ventral hippocampus in vivo. Behav Pharmacol 1994; 5:299-305.

82. Adell A, Casanovas JM, Artigas F. Comparative study in the rat of the actions of different types of stress on the release of 5-HT in raphe nuclei and forebrain areas. Neuropharmacology 1997; 36:735-741.

83. Barnes NM, Cheng CHK, Costall B, Ge, J., Kelly, M.E., Naylor, R.J. Profiles of interaction of R(+)/S(-)zacopride and anxiolytic agents in a mouse model. Eur J Pharmacol 1992a; 218:91-100.

84. Barnes NM, Costall B, Ge J, Kelly ME, Naylor RJ. The interaction of R(+)- and S(-)-zacopride with PCPA to modify rodent aversive behavior. Eur J Pharmacol 1992c; 218:15-25.

85. Ge J, Barnes NM. 5-HT₄ receptor-mediated modulation of 5-HT release in the rat hippocampus in vivo. Br J Pharmacol 1996; 117:1475-1480.

86. Silvestre JS, Fernández AG, Palacios JM. Effects of 5-HT₄ receptor antagonists on rat behavior in the elevated plus-maze test. Eur J Pharmacol 1996; 309:219-222.

87. Iversen SD. 5-HT and anxiety. Neuropharmacology 1984; 23:1553-1560

88. Handley SL, McBlane JW. 5-HT drugs in animal models of anxiety. Psychopharmacology 1993; 112:13-20

89. Costall B, Naylor RJ. The pharmacology of the 5-HT₄ receptor. Int Clin Psychopharm 1993; 8(suppl 2):11-18.

90. Cheng CHK, Costall B, Kelly ME, Naylor RJ. Actions of 5-hydroxytryptophan to inhibit and disinhibit mouse behavior in the light/dark test. Eur J Pharmacol 1994; 255:39-49.

91. Costall B, Naylor RJ. 5-HT₄ receptor antagonists attenuate the disinhibitory effects of diazepam in the mouse light/dark test. Br J Pharmacol 1996; 119(suppl):352P.

92. Grossman CJ, Kilpatrick GJ, Bunce KT. Development of a radioligand binding assay for 5-HT₄ receptors in guinea-pig and rat brain. Br J Pharmacol 1993; 109:618-624.

93. Jakeman LB, To ZP, Eglen RM, Wong EHF, Bonhaus DW. Quantitative autoradiography of 5-HT₄ receptors in brains of three species using two structurally distinct radioligands, [³H]GR113808 and [³H]BIMU-1. Neuropharmacology 1994; 33:1027-1038.

CHAPTER 6

5-Hydroxytryptamine and Human Heart Function: The Role of 5-HT$_4$ Receptors

Alberto J. Kaumann and Louise Sanders

Introduction

5-Hydroxytryptamine (5-HT) can elicit both cardioexcitation and cardiodepression. 5-HT exerts these effects both through a direct action in the heart and indirectly via stimulation of the central nervous system (CNS). Here we concentrate mainly on the direct effects of 5-HT on the heart.

Cardioexcitation

Cardioexcitation, such as increased beating rate and force, is mediated by a vast array of 5-HT receptors that depend on the species. At the lower end of the evolutionary spectrum, 5-HT is an excitatory neurotransmitter, acting in mollusk hearts in a cyclic AMP-dependent manner[1] through unknown 5-HT receptors.[1,2] In mammals the 5-HT-induced cardioexcitation is mediated through 5-HT$_{1\text{-like}}$ receptors in the cat[3,4] (possibly 5-HT$_7$),[5] 5-HT$_{2A}$ receptors in the rat,[6] 5-HT$_3$ receptors in the rabbit (through release of noradrenaline),[7] possibly both 5-HT$_3$ and 5-HT$_4$ receptors in the guinea pig,[8] and 5-HT$_4$ receptors in pig[9-12] and man.[13-23] The cardioexcitatory receptors are usually located on the myocardial cells. Exceptions are 5-HT$_3$ receptors located on sympathetic nerve terminals in rabbit heart[7] and possibly in guinea-pig atria.[8] In mammals the 5-HT-evoked cardioexcitation is observed primarily in atrial tissue, both in the myocardium and the sinoatrial node.[3,4,8,9] In ventricular tissue contractile force is enhanced by high concentrations of 5-HT in the cat[3,4] and guinea pig[23] through prolongation of the action potential, but it is not known if or which 5-HT receptors are involved.

Cardiodepression

Cardiodepression occurs indirectly in many species at the peripheral level through excitation of sensory afferent vagal nerve endings in the heart via activation of 5-HT$_3$ receptors.[7,24] This initiates the Bezold-Jarisch reflex, leading to a slowing of heart rate. In man 5-HT may also potentially depress heart function indirectly by constricting large coronary arteries through activation of both 5-HT$_{1-like}$ and 5-HT$_{2A}$ receptors.[25-28] The 5-HT$_{1-like}$ receptors that mediate constriction of human large coronary arteries are likely to correspond to the cloned 5-HT$_{1B}$ receptors.[28] The ensuing arterial spasm may cause myocardial ischemia with subsequent depression of heart function in some patients (discussed in refs. 27 and 28).

5-HT$_4$ Receptors in the Human Heart

As mentioned above and unlike most other species, 5-HT increases human atrial contraction through 5-HT$_4$ receptors.[13-23] We discuss below the properties of human atrial 5-HT$_4$ receptors and how these compare to those of CNS and other peripheral 5-HT$_4$ receptors as well as recombinant 5-HT$_4$ receptors. We also discuss some aspects of the functional modulation of human atrial 5-HT$_4$ receptors. We propose that endogenous 5-HT may reach atrial 5-HT$_4$ receptors not only through platelet aggregation but also indirectly through release from cardiac sympathetic nerve endings. We discuss an experimental model demonstrating the generation of 5-HT-evoked atrial arrhythmias[28] through 5-HT$_4$ receptors in vitro and the possible clinical relevance of this finding. We suggest that 5-HT$_4$ receptors mediate the tachycardia evoked by intravenous administration of 5-HT to humans. We consider the plausible participation of cardiac vagal 5-HT$_4$ receptors in 5-HT-induced bradycardia. We also briefly address the effects of 5-HT on ventricular tissue.

Properties

It was noticed in 1983 that 5-HT causes concentration-dependent increases in the force of contraction of isolated human right atrium obtained from patients without advanced heart failure (A. J. Kaumann, unpublished experiments). No available 5-HT receptor antagonist blocked this effect and the nature of the 5-HT receptor involved remained unknown until 1989. Then, renewed interest and an expanded knowledge of the ever-increasing number of subtypes of 5-HT receptors, together with the consequent increase in the armory of 5-HT receptor subtype-selective ligands, enabled the identification of the receptor subtype involved.

5-HT was shown to be five times more potent than (-)-noradrenaline[13,14] as an inotropic agonist with a maximum effect of around 60% of that of a saturating concentration of catecholamines in studies using right atrial strips from β-adrenoceptor blocker-treated patients.[13,14,21] The 5-HT-evoked positive inotropic effects proved to be resistant to the in vitro blockade of β-adrenoceptors and both β_1- and β_2-adrenoceptors, thus ruling out indirect increases in contractility through a 5-HT-induced release of noradrenaline, e.g., from cardiac nerve endings.[13,21] The 5-HT-evoked positive inotropic effects were also resistant to blockade by antagonists of 5-HT_{1A}, 5-HT_{1B}, 5-HT_{1D}, $5\text{-HT}_{1\text{-like}}$, 5-HT_{2A} and 5-HT_{2C} receptors and by the 5-HT_3 receptor-selective antagonists MDL 72222 and granisetron, but were competitively blocked with moderate potency by the 5-HT_3 receptor-selective antagonist tropisetron (ICS 205-930) (which also blocks 5-HT_4 receptors with moderate potency)[13] and by the partial agonists renzapride and cisapride.[14] In addition, 5-HT and, to a lesser extent renzapride and cisapride, were shown to cause increases in cyclic AMP levels and cyclic AMP-dependent protein kinase (PKA) activity in human atrial tissue.[13,14,16,21] The receptors mediating the 5-HT-evoked positive inotropic effects were therefore proposed[13,14,29] to resemble tropisetron-sensitive 5-HT_4 receptors in rodent brain,[30-34] which stimulate adenylyl cyclase.

Our initial observations of the positive inotropic effects of 5-HT and stimulation of the cardiac cyclic AMP-dependent pathway through 5-HT_4 receptors have been confirmed by several laboratories.[35-39]

The 5-HT_4 nature of human atrial 5-HT receptors received considerable support with the introduction of the 5-HT_4 receptor-selective antagonists SB 203186, GR 113808 and SB 207710, which block these receptors competitively with high affinity (Table 6.1). The most potent of these compounds, SB 207710, contains an iodine atom and has been used successfully as a radioiodinated probe to label atrial 5-HT_4 receptors.[12,22] The density of 5-HT_4 receptors is around 10% and 20% of the densities of β_1- and β_2-adrenoceptors, respectively in

Table 6.1. Affinity and potency of ligands for human atrial 5-HT$_4$ receptors

	pK$_B$	pK$_D$	pEC$_{50}$	pK$_P$	Refs.
[^{125}I] SB 207710		9.6			22
SB 207710	10.1	9.7			19, 22
GR 113808	8.8				18
SB 203186	8.7	8.0			11, 22
SDZ 205-557	7.7				39
Tropisetron	6.7	6.1			13, 22
5-HT		5.8	7.4-7.9		13, 14, 21, 22
5-CT		4.9	4.7		14, 22
Renzapride		6.4	6.3	6.7	14, 22
Cisapride		6.0	6.1	6.2	14, 22

pK$_B$, pK$_P$ and pK$_D$ are equilibrium dissociation constants (-log M), estimated from antagonism of the effects of 5-HT by an antagonist (B) or partial agonist (P), and from binding (D) of ligands to 5-HT$_4$ receptors. pEC$_{50}$ is -log M of the EC$_{50}$ for agonists and partial agonists.

human atrium.[22] Incidentally, in piglet atria the density of 5-HT$_4$ receptors is less than 1% of that of β_1-adrenoceptors.[12] The relatively low 5-HT$_4$ receptor densities in human and piglet atrium may account for the smaller maximum positive inotropic effects of 5-HT in human[13,14,16,21] and porcine atrium[40,41] compared to those evoked by catecholamines through β-adrenoceptors.[12,22] The very low density of piglet atrial 5-HT$_4$ receptors compared to human atrial 5-HT$_4$ receptors is reflected by smaller 5-HT-evoked contractions in porcine atria[40,41] than in human atria.[13,14,16,21] In human atrial myocytes, however, 5-HT causes the same maximum contractile shortening as (-)-isoprenaline and 5-HT is also more potent than in atrial tissues.[21] The more efficacious and potent effects of 5-HT on atrial myocytes compared to tissues could conceivably be due to the appearance of additional 5-HT$_4$ receptors uncovered by enzymes used to disaggregate the cells from atrial tissue.

In addition to enhancing atrial contractile force 5-HT hastens the relaxation (lusitropic effect) of the 5-HT-stimulated muscle.[13] Positive inotropic and lusitropic effects similar to those of 5-HT are also obtained with (-)-noradrenaline and (-)-adrenaline in human right atrial appendage tissue in vitro.[42] Both catecholamines also cause increases in cyclic AMP levels and PKA activity in the atrium[43] and the involvement of the cyclic AMP-dependent pathway in the production of the inotropic and lusitropic effects seems likely. By analogy with the effects of the catecholamines, 5-HT-induced increases of cyclic AMP levels and PKA activity occur concominantly with the positive inotropic and lusitropic effects of 5-HT,[13,14,21] suggesting the participation of the cyclic AMP-dependent cascade in producing these responses.[13,14] It has been proposed[13,14] that the 5-HT-evoked PKA activity results in the phosphorylation of the cardiac sarcolemmal L-type Ca^{2+} channel, phospholamban and troponin I. Phosphorylation of the L-type Ca^{2+} channel enhances its permeability for Ca^{2+} thereby facilitating Ca^{2+}-stimulated Ca^{2+} release from the sarcoplasmic reticulum (SR) and thus accounts for the positive inotropic effects of 5-HT. Phosphorylation of phospholamban on the SR removes the inhibitory effect of nonphosphorylated phospholamban on the Ca^{2+}-ATPase pump thereby enhancing its Ca^{2+} pumping activity. As a result sarcoplasmic Ca^{2+} concentrations are reduced with ensuing relaxation of contractile proteins, and more Ca^{2+} is available for release from the sarcoplasmic reticulum. Phosphorylation of troponin I reduces the affinity of troponin C for Ca^{2+}, thus also facilitating relaxation.

Further evidence for the involvement of the cyclic AMP-dependent cascade in the 5-HT-stimulated inotropic and lusitropic effects comes from electrophysiological data. 5-HT causes a 6- to 7-fold increase in peak L-type Ca^{2+} channel current in human atrial myocytes.[15,35,36] The effect is blocked by an inhibitor of PKA and is nonadditive with the effects of intracellularly administered cyclic AMP, consistent with an obligatory participation of PKA and phosphorylation of L-type Ca^{2+} channels.[15] The 5-HT-induced increase in Ca^{2+} current is due to a greater availability of L-type Ca^{2+} channels, probably related to PKA-induced channel phosphorylation.[15] Mean open and shut times, as well as the open probability of a single channel are unaffected by 5-HT. The enhanced L-type Ca^{2+} channel current elicited by 5-HT is readily abolished by washout of 5-HT.[36]

Comparison with Noncardiac and Recombinant 5-HT₄ Receptors

Although the human atrial 5-HT_4 receptor greatly resembles the 5-HT_4 receptor identified in rodent brain,[30-34] there are some striking differences in the interaction of certain compounds at these two receptor groups.[14] As also shown for the piglet sinoatrial node,[9,44] the substituted benzamides renzapride and cisapride are partial agonists of low potency and efficacy as positive inotropic agents in human right atrium compared to 5-HT,[14] whereas they are as potent as and actually more efficacious than 5-HT at stimulating adenylyl cyclase activity in mouse embryonic colliculi neurones in culture.[30] It seems, therefore, that human and piglet 5-HT_4 receptors are not necessarily identical to mouse cerebral 5-HT_4 receptors; for this reason the atrial 5-HT receptors have sometimes been referred to as $5\text{-HT}_{4\text{-like}}$ receptors.[8,9,14,16] The pharmacology of human atrial[13,14,16] and porcine atrial[9,40,41,44] 5-HT_4 receptors appears, however, similar to that of 5-HT_4 receptors found on the cholinergic neurones of guinea-pig ileum[45] and colon,[46] where they activate contraction, and the smooth muscle of rat esophagus[47,48] and human colon[49] where they mediate relaxation. Can the pharmacological differences between central and some peripheral 5-HT_4 receptors be explained with the recent identification of two splice variants of recombinant 5-HT_4 receptors, short form ($5\text{-HT}_{4\text{S}}$) and a long form ($5\text{-HT}_{4\text{L}}$), or are they merely the result of species homologues?

The rat,[50] human[51] and mouse[52] 5-HT_4 receptors have been cloned and $5\text{-HT}_{4\text{L}}$ and $5\text{-HT}_{4\text{S}}$ splice variants detected for each of these species. Both splice variants are expressed in rat[50] and mouse[52] brain regions, but only mRNA for the short splice variant has been found in rat atrium.[50] The functional relevance of rat atrial $5\text{-HT}_{4\text{S}}$ mRNA is obscure because there is no pharmacological evidence for rat atrial 5-HT_4 receptors[8] and 5-HT-evoked tachycardia in the rat appears to be mediated through $5\text{-HT}_{2\text{A}}$ receptors.[6] Perhaps the $5\text{-HT}_{4\text{S}}$ mRNA was located in rat atrial endothelial or endocardial cells because 5-HT_4 receptor mRNA has been reported to be expressed in human endothelial cells.[53] Mouse cerebral and recombinant $5\text{-HT}_{4\text{L}}$ receptors[52] have higher binding affinities and efficacy for cisapride and renzapride than the human atrial 5-HT_4 receptors. Also, cisapride is as potent[30] or even more potent than renzapride[52] in activating them while the opposite was found for 5-HT_4 receptors in human atrium.[14] Recently a human atrial 5-HT_4 receptor, $h5\text{-HT}_{4\text{A}}$, has been cloned that resembles the short splice variant of the rat 5-HT_4 receptor.[51] $h5\text{-HT}_{4\text{A}}$ mRNA was detected in human atrium but not in ventricle.[51] The pharmacology of the transfected $h5\text{-HT}_{4\text{A}}$ receptor, as assessed in COS cells, is consistent with that of human and piglet atrial $5\text{-HT}_{4\text{-like}}$ receptors.[9,12,14,22] The most striking results are the low affinity and agonist efficacy of renzapride at the recombinant $h5\text{-HT}_{4\text{A}}$ receptors[51] as found with the native human and piglet atrial receptors.[9,12,14,22] This evidence suggests that the human atrial $5-HT_{4-like}$ receptor corresponds to the recombinant $h5-HT_{4A}$ receptor which may be a homologue of the rat $5-HT_{4S}$ receptor.

Modulation of Human Atrial 5-HT$_4$ Receptors

Human atrial $5-HT_4$ receptors that mediate increases in contractile force appear to undergo surprisingly little desensitization upon acute exposure to $5-HT$[13,14,16,17,21] compared to $5-HT_4$ receptors of embryonic mouse colliculi neurones.[32] 5-HT can cause positive inotropic effects ex vivo in several regions of human left atrium obtained from patients with terminal heart failure by enhancing cyclic AMP levels and stimulating PKA activity.[16] Moreover, the inotropic efficacy and potency of 5-HT in left atrial tissues from patients with terminal heart failure are similar to those in right atrial tissues from nonfailing hearts obtained from patients not treated with β blockers.[16,21] These experiments suggest that $5-HT_4$ receptors in atria from failing hearts undergo little desensitizaton. Possible reasons for the marked desensitization of murine neuronal $5-HT_4$ receptors compared to the relatively small desensitization of human atrial $5-HT_4$ receptors could be the expression of different splice variants of the 5-HT receptor in different tissues or differences in the species homologues of $5-HT_4$ receptors.

When increases in L-type current are measured in atrial cardiomyocytes obtained from hearts in terminal failure, 5-HT-evoked increases in L-type Ca^{2+} current appear, however, to be smaller than in cardiomyocytes from nonfailing hearts.[54] A similar decrease of isoprenaline-induced stimulation of L-type Ca^{2+} current but also a decrease in basal Ca^{2+} channel density has also been observed in ventricular myocytes from human hearts in terminal failure compared to myocytes from nonfailing hearts.[54,55] These findings suggest a nonspecific desensitization of receptor-mediated responses, perhaps through blunting of the function of some common step of the cyclic AMP-dependent cascade downstream from the receptors.[54]

Hyperfunction of human atrial $5-HT_4$ receptors has been detected in tissues and cardiomyocytes from the hearts of patients chronically treated with β blockers usually selective for $β_1$-adrenoceptors.[21] 5-HT becomes a more efficacious and potent inotropic agonist in atrial tissues and myocytes from β blocker-treated patients compared to non-β blocker-treated patients. This property is not specific for $5\text{-}HT_4$ receptors, because hyperresponsiveness of other G_s protein-coupled receptors has been reported, namely for β_2-adrenoceptors,[42,43] histamine H_2 receptors,[56] β_1-adrenoceptors[43,57] and (only marginally) putative β_4-adrenoceptors.[58,59] The receptor hyperresponsiveness is a function of receptor type with the following rank order:

$$\beta_2\text{-adrenoceptor} > 5\text{-}HT_4 \text{ receptor} \cong H_2 \text{ receptor} >> \\ \beta_1\text{-adrenoceptor} > \beta_4\text{-adrenoceptor}$$

There is not yet a unifying theory to account for the receptor hyperresponsiveness caused by chronic treatment of patients with β blockers. One reason is that biochemical information about receptor cross-talk in human atrium is scarce. Both $5\text{-}HT_4$ receptors[21] and histamine H_2 receptors[56] mediate ex vivo greater increases in cyclic AMP levels in atria from β blocker-treated patients than in atria from non-β blocker-treated patients. On the other hand, the positive inotropic responses to dibutyryl cyclic AMP[42] and forskolin[21] are not increased by chronic β blocker treatment in atrial tissues and myocytes respectively, apparently ruling out the possibility that the inotropic hyperresponsiveness is due to a modification of effectors downstream of cyclic AMP. The densities of β_1- and β_2-adrenoceptors do not differ in atria obtained from patients chronically treated or not treated with β blockers as assayed using two independent methods[57,60] but nothing is currently known about possible changes in densities of $5\text{-}HT_4$ receptors or histamine H_2 receptors. Taken together, the incomplete evidence so far suggests an increased coupling of several receptor types to G_s protein after chronic prevention of the effects of noradrenaline released from sympathetic nerves through blockade of β_1-adrenoceptors. One possibility is that increased coupling occurs through phosphorylation of G_s protein by a tyrosine protein kinase, which enhances G_s protein function,[61] but there is not yet evidence for this. Tonic activation of β_1-adrenoceptors by noradrenaline would thus blunt the coupling of several receptor types to G_s protein through unknown mechanisms. It is likely that more than one mechanism is involved. For example, in the case of the hyperresponsiveness of atria from β blocker-treated patients to histamine, not only H_2 receptors are involved but also H_1 receptors. H_1 receptors mediate a marked increase in atrial cyclic GMP levels, which in turn enhance cyclic AMP levels and PKA activity, presumably by inhibiting the hydrolysis of cyclic AMP through phosphodiesterase III (an enzyme inhibited by cyclic GMP), and this effect appears to cause, in part, the inotropic hyperresponsiveness to histamine.[56] Another factor that may contribute to the receptor-selective hyperresponsiveness caused by chronic β blocker treatment is receptor-dependent tightness of coupling to G_s protein. $β_2$-adrenoceptors are more tightly coupled to G_s protein than $β_1$-adrenoceptors[62-65] but comparative information about the relative coupling of $5-HT_4$ receptors, histamine H_2 receptors and putative $β_4$-adrenoceptors is not yet available.

Endogenous 5-HT

How could 5-HT affect the cardiovascular system in vivo? What is the source of the 5-HT that could cause these cardiovascular effects? How could endogenous 5-HT reach the heart? Four possible sources of 5-HT are relevant: the heart itself, nerve endings, mast cells and platelets. The human heart contains around 400 ng 5-HT per g tissue; its histological localization, synthesis, metabolism, release mechanism and function are unknown.[66] The capture of platelet-derived 5-HT by sympathetic nerves and its subsequent release has been demonstrated for canine coronary arteries.[67] It is feasible that sympathetic nerve endings of human atrium can capture 5-HT, which could then be released by electrical activation of the nerves to interact with $5-HT_4$ receptors and thereby enhance contractile force. This is demonstrated in vitro with the experiments of Figure 6.1.

The most obvious potential source of 5-HT acting on the human heart is the platelets. It has been suggested that under normal conditions the endocardium protects the heart by preventing platelet aggregation, but that when the endocardium is damaged, platelet aggregation and the associated release of 5-HT may be facilitated.[68] Is the effect of endogenously released 5-HT beneficial or harmful to the heart? There is no evidence for beneficial effects. A potentially harmful effect could be the ability of 5-HT to evoke arrhythmias in human atrium. We have studied this property in an experimental model.

Experimental 5-HT-Evoked Arrhythmias

While investigating the positive inotropic effects of 5-HT in paced human atrial tissues[13,14,17,21] and myocytes,[21] we noticed the occasional appearance of arrhythmias, consisting of extrasystoles and/or pacemaker activity. The 5-HT-induced arrhythmias are

Fig. 6.1. Release of 5-HT
by field stimulation and
interaction with 5-HT$_4$
receptors. Results from
four trabeculae of a
right atrial appendage
from a 50 year old male
patient undergoing
coronary artery sur-
gery. The experiment
was carried out at 37°C
in the presence of 200
M ascorbate. The trabe-
culae were bathed in
modified Krebs solu-
tion and paced at 1 Hz
through a punctiform
electrode. To release
neurotransmitters, the
tissues were stimulated
with field stimuli deliv-
ered through field elec-
trodes into the absolute
refractory period of the
cardiac action potential
(without re-exciting the
trabeculum) at the in-
dicated frequency (Hz)

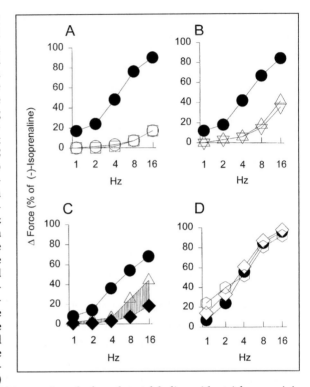

as described.[101] To avoid interaction of released acetylcholine with atrial muscarinic
receptors, all experiments were carried out in the presence of 1 M atropine. Three suc-
cessive curves relating field stimulation frequency in Hz to contractile force were determined on each trabeculum. Increases of contractile force were expressed as a percentage of the increase in peak contractile force caused by 200 M (-)-isoprenaline, administered at the end of the experiment. Each panel represents results from a single
trabeculum. The first field stimulation-force curve, determined in the absence of any
antagonist, is shown by filled circles. In A the second (open circles) and third (open
squares) curves were determined in the presence of 200 nM (-)-propranolol, incubated
for 45 min before the second curve was begun. In B the experiments for the second and
third curves were carried out in the presence of 200 nM (-)-propranolol as in A but the
tissues were incubated for 30 min with 10 M 5-HT followed by 10 min washout before
determination of the second (open triangles) and third curve (inverted open triangles).
The protocol of the experiment in C was as in B, except that the 5-HT$_4$ receptor-selective antagonist SB 207710 (100 nM)[19] was added after the second curve was finished
and was present during the third curve (filled diamonds). In D no 5-HT was present
and the second curve (open diamonds) was determined in the absence of SB 207710,
the third curve (open hexagons) in the presence of SB 207710 (100 nM). The hatched
area in C represents the increase in contractile force elicited from neuronally captured
and released 5-HT interacting with 5-HT$_4$ receptors. Similar results were obtained from
another two experiments. Unpublished experiments of A. J. Kaumann and P. Molenaar.

presumably related to Ca^{2+} overload resulting from $5\text{-}HT_4$ receptor-mediated activation of PKA that phosphorylates the L-type Ca^{2+} channel, thereby enhancing its permeability.[15,20,36] In addition, 5-HT increases the pacemaker current i_f[69] through $5\text{-}HT_4$ receptors in human atrium[70] and this may further facilitate arrhythmias. The involvement of $5\text{-}HT_4$ receptors is supported by the finding that the arrhythmias are blocked by the selective $5\text{-}HT_4$ receptor antagonist SB 203186.[17] The incidence of 5-HT-evoked arrhythmias is an inverse function of heart rate (i.e., atrial pacing rate).[17] As expected from the $5\text{-}HT_4$ receptor hyperresponsiveness in atria obtained from β blocker-treated patients,[21] the incidence of 5-HT-evoked arrhythmias in these tissues is also higher than that of atria from non-β blocker-treated patients.[17] These experiments support the proposal that platelet-derived 5-HT may cause atrial fibrillation through $5\text{-}HT_4$ receptors.[20]

Hypothesis: 5-HT Causes Atrial Fibrillation Through 5-HT₄ Receptors

Atrial fibrillation is the most frequently occurring type of arrhythmia with concomitant deleterious effects, and its incidence increases with age.[71,72] With increasing age hemodynamic dysfunction due to valvular and nonvalvular heart disease cause left atrial enlargement which may facilitate slow conduction, a requirement for fibrillation. Left atrial size contributes to the incidence[73-75] and duration[75] of atrial fibrillation. Once established, atrial fibrillation in turn appears to enlarge the left atrium.[76,77] As suggested for ventricular endocardium,[68] the left atrial endocardium may become damaged; lesions have been demonstrated, for example, in rheumatic heart disease.[78] There is also evidence for stagnant blood flow in the enlarged left atrium of patients with mitral valve disease, two thirds of whom have atrial fibrillation.[79] The disruption of the atrial endocardium may facilitate platelet adhesion and aggregation, leading to the release of high local concentrations of 5-HT and other platelet-derived vasoactive substances. The sluggish blood flow, together with the lesions of the endocardium may contribute to the platelet aggregation and formation of thrombi. Multiple thrombi have been found in human left atria with atrial fibrillation but without valvular disease,[80] and it is well known that these thrombi are the main source of cerebral embolism and stroke.[81,82] The risk of stroke is considerably increased by atrial fibrillation: 17-fold in patients with rheumatic heart disease and 5-fold in patients with nonrheumatic heart disease.[82] The risk of stroke is highest after the onset of atrial fibrillation.[83] The mechanism of induction of atrial fibrillation and stroke is conceivably as follows: 5-HT released from aggregating platelets activates $5\text{-}HT_4$ receptors, thereby inducing arrhythmias, followed by atrial fibrillation. The massive release of 5-HT that occurs during platelet aggregation enhances atrial contractions, thereby dislodging parts of the thrombus and causing emboli which in turn produce stroke. The dislodgement of the thrombus may be further facilitated by the uncoordinated contractions due to the atrial fibrillation. We have demonstrated both the positive inotropic[13,14,21,23] and proarrhythmic[17] (and unpublished work from our laboratory) effects of 5-HT in vitro, thus supporting the proposed mechanism. Because $5\text{-}HT_4$ receptors mediate these effects[13,14,17,21] a selective $5\text{-}HT_4$ receptor antagonist would be expected to prevent atrial fibrillation, thereby reducing the incidence of stroke.[20] Our evidence showing that a $5\text{-}HT_4$ receptor-selective antagonist blocks 5-HT-evoked atrial arrhythmias in vitro is consistent with, but does not prove, the above prediction. Ultimate support or rejection of the hypothesis that $5\text{-}HT_4$ receptors can be involved in the etiology of atrial fibrillation and stroke can only come from large-scale studies of the long-term effects of chronic administration of $5\text{-}HT_4$ receptor antagonists in the elderly, once these compounds are available for clinical investigation.

Some circumstantial evidence already exists in support of the hypothesis that $5\text{-}HT_4$ receptors mediate atrial fibrillation. Patients with carcinoid tumors and carcinoid heart disease have high blood levels of vasoactive substances, including 5-HT, and can occasionally exhibit atrial fibrillation or ectopic atrial rhythm.[84] Since the incidence of right-sided heart lesions in patients with carcinoid tumors is much greater than the incidence of left-sided lesions,[84,85] it is probable that the vasoactive substances secreted by the tumor are cleared in the lungs. Through this mechanism the level of free 5-HT reaching the left atrium may be low, perhaps explaining why patients with carcinoid tumors do not have a high incidence of atrial arrhythmias.

It has recently been reported that cisapride can cause tachycardia and generate supraventricular arrhythmias in man.[86,87] Because cisapride is a partial agonist at human atrial $5\text{-}HT_4$ receptors,[14] it is likely that cisapride-evoked arrhythmias are also mediated through $5\text{-}HT_4$ receptors. The evidence with cisapride supports the concept that $5\text{-}HT_4$ receptor activation in human atrium is arrhythmogenic.

The Effects of 5-HT on Human Heart Rate

Studies carried out in the late 1950s showed that 5-HT, administered intravenously, can cause both a decrease[88] and an increase[89-91] in heart rate in man. The bradycardia preceded the tachycardia in some individuals[88] but the most consistent effect of 5-HT was tachycardia[88-91] which was concentration-dependent.[90] 5-HT also increases respiration and forearm blood flow but these effects are preceded by the initiation of tachycardia,[90,91] suggesting a direct interaction of 5-HT with the sinoatrial node.

Bradycardia

The acute drop in heart rate sometimes observed in man upon the intravenous administration of 5-HT is accompanied by hypotension.[88] Similar effects have been seen in a variety of other species.[7,24] In animal species, the 5-HT-evoked bradycardia is prevented by vagotomy, atropine and the 5-HT_3 receptor antagonist MDL 72222,[7,24] consistent with the involvement of 5-HT_3 receptors located on afferent vagal fibers. Interaction of 5-HT with these 5-HT_3 receptors is followed by activation of efferent vagal fibers and subsequent release of acetylcholine that causes bradycardia through sinoatrial muscarinic receptors—the Bezold-Jarisch reflex.[7,24] The observed hypotension[24] is due to decreased arteriolar constriction of neural origin arising from decreased sympathetic activity, and probably also to a decrease in cardiac output. Direct evidence verifying the existence of the 5-HT-induced Bezold-Jarisch reflex in man is still lacking.

In the rat, depolarization of the vagus is mediated by two types of 5-HT receptors: 5-HT_3 receptors, as above, and a receptor that is depolarized by 5-HT_4 receptor agonists and blocked by the weak 5-HT_4 receptor antagonist tropisetron, suggesting the coexistence and cofunction of both 5-HT_3 and 5-HT_4 receptors.[92] Whether or not activation of these 5-HT_4 receptors contributes to the Bezold-Jarisch reflex in the rat, and whether or not they even exist on the human vagus, is not yet known.

Tachycardia

Which 5-HT receptor mediates 5-HT-induced tachycardia in man? Although conclusive data are still lacking, the following evidence supports the hypothesis that 5-HT-induced tachycardia is mediated by 5-HT_4 receptors located on the sinoatrial node. 5-HT_4 receptors have been identified in isolated human right and left atrium and localized in human atrial myocytes.[13-22] 5-HT evokes tachycardia in the pig through receptors that greatly resemble human atrial 5-HT_4 receptors.[9-12,44,93] In isolated tissues, piglet sinoatrial[9,44] and human right atrial[13,14] 5-HT_4 receptors have the same order of potency for agonists and partial agonists

$$5\text{-HT} > \text{renzapride} > \text{cisapride} \gg$$
$$5\text{-carboxamido-tryptamine (5-CT)},$$

similar intrinsic activities for the partial agonists renzapride and cisapride relative to the activity of 5-HT and a similar order of affinities for 5-HT_4 receptor antagonists (Table 6.1).

$$\text{SB 207710} > \text{GR 113808} \cong \text{SB 203186} > \text{SDZ 205-557} > \text{tropisetron}$$

The existence of porcine sinoatrial 5-HT_4 receptors[9,44] has been verified in intact adult anesthetized pigs[10,11,93] and the 5-HT_4 receptor agonist rank order of potency found in isolated tissues of man[14] and piglet[9,44] is replicated in intact adult pigs.[10,93] Furthermore, both tropisetron[10] and SB 203186[11] antagonize 5-HT-evoked tachycardia in intact adult anesthetized pigs, with SB 203186 being considerably more potent an antagonist than is tropisetron,[11] as found in vitro.[11,44]

Cisapride, a weak partial agonist on both human right atrium[14] and porcine sinoatrial node[9] (as discussed above) can also cause tachycardia in man[84,85,94] and it has been suggested that this occurs through activation of sinoatrial 5-HT_4 receptors.[95,96] Conceivably, cisapride could cause hypotension and the cisapride-evoked tachycardia is due to a reflex release of (-)-noradrenaline and subsequent activation of sinoatrial β-adrenoceptors. This is, however, unlikely because 8 mg cisapride administered intravenously in healthy volunteers evoked tachycardia without altering mean blood pressure, although pulse pressure was slightly increased.[94]

A direct way of testing the hypothesis that the tachycardia evoked by 5-HT and cisapride in man is mediated directly through sinoatrial 5-HT_4 receptors would be to administer 5-HT or cisapride in a cumulative fashion directly to the sinoatrial node in healthy volunteers. To avoid leakage into the peripheral circulation and the resulting possible elicitation of reflexes, the 5-HT or cisapride could be injected in small amounts through a catheter placed into a branch of the coronary artery that supplies the sinoatrial node. If 5-HT_4 receptors were involved in any resulting tachycardia evoked by 5-HT or cisapride, a 5-HT_4 receptor-selective antagonist would be expected to block the tachycardia. To rule out the participation of endogenous catecholamines and activation of sinoatrial β-adrenoceptors, the volunteers should also take a nonselective blocker of β_1- and β_2-adrenoceptors.

5-HT and the Human Ventricle

Following the discovery of human atrial 5-HT_4 receptors, it was suggested that 5-HT and related compounds may serve as inotropic agents in cases of severe heart failure when β-adrenoceptor function is depressed.[15,97] However, no evidence for 5-HT_4 receptors has been found in human ventricle.[35,38] 5-HT does not enhance contractile force of isolated ventricular preparations[38] nor does it affect L-type Ca^{2+} channel current in human ventricular myocytes.[35]

As reviewed,[98] 5-HT appears to cause ventricular arrhythmias through 5-HT_{2A} receptors in the rat. There is no evidence for 5-HT_{2A} receptors that mediate arrhythmias in human myocardium. Conceivably, however, 5-HT may indirectly elicit ventricular and even supraventricular (atrial) arrhythmias through induction of coronary artery spasm via 5-HT_{2A} receptors, as suggested.[98] Because $5\text{-HT}_{1\text{-like}}$ receptors (most likely 5-HT_{1B} receptors) also participate in the mediation of 5-HT-induced contraction of large human coronary arteries,[28] these receptors may also mediate coronary spasm. Evidence for coronary spasm[99] and myocardial infarct[100] that appeared following administration of the $5\text{-HT}_{1\text{-like}}$ receptor agonist sumatriptan suggests involvement of 5-HT_{1B} receptors.[28]

Acknowledgment

AJK is grateful to the British Heart Foundation for support.

References

1. Sawada M, Ichinose M, Ito I, Maeno T, Mcadoo DJ. Effects of 5-hydroxytryptamine on membrane potential, contractility, accumulation of cyclic AMP, and Ca^{2+} movements in anterior aorta and ventricle of aplysia. J Neurophysiol 1984; 51:361-374.
2. Greenberg MJ. Structure-activity relationshp of tryptamine analogues on the heart of venus mercenaria. Br J Pharmacol 1960; 15:375-388.
3. Kaumann AJ. Two classes of myocardial 5-hydroxytryptamine receptors that are neither 5-HT_1 nor 5-HT_2. J Cardiovasc Pharmacol 1985; 7(Suppl):S76-S78.
4. Kaumann AJ. Further differences between 5-HT receptors of atrium and ventricle in cat heart. Br J Pharmacol 1986; 89:546 P.
5. Villalon CM, Heiligers JPC, Centurion D, DeVries P, Saxena PR. Characterization of putative 5-HT_7 receptors mediating tachycardia in the cat. Br J Pharmacol 1997; 121:1187-1195.

6. Docherty JR. Investigations of cardiovascular 5-hydroxytryptamine receptor subtypes in the rat. Naunyn-Schmiedeberg's Arch Pharmacol 1988; 337:1-8.

7. Fozard JR. MDL 72222, a potent and highly selective antagonist at neuronal 5-hydroxytryptamine receptors. Naunyn-Schmiedeberg's Arch Pharmacol 1984; 326:36-44.

8. Kaumann AJ. 5-HT$_{4\text{-like}}$ receptors in mammalian atria. J Neural Transm 1991; 34(Suppl):195-201.

9. Kaumann AJ. Piglet sinoatrial 5-HT receptors resemble human atrial 5-HT$_{4\text{-like}}$ receptors. Naunyn-Schmiedeberg's Arch Pharmacol 1990; 342:619-622.

10. Villalon CM, DenBoer MO, Heiligers JPC, Saxena PR. Mediation of 5-hydroxytryptamine-induced tachycardia in the pig by the putative 5-HT$_4$ receptor. Br J Pharmacol 1990; 100:665-667.

11. Parker SG, Taylor EM, Hamburger SA, Vimal M, Kaumann AJ. Blockade of human and porcine myocardial 5-HT$_4$ receptors by SB 203186. Naunyn-Schmiedeberg's Arch Pharmacol 1995; 335:28-35.

12. Kaumann AJ, Lynham JA, Brown AM. Labelling with [^{125}I]-SB 207710 of a small 5-HT$_4$ receptor population in piglet right atrium: functional relevance. Br J Pharmacol 1995; 115:933-936.

13. Kaumann AJ, Sanders L, Brown AM, Murray KJ, Brown MJ. A 5-hydroxytryptamine receptor in human atrium. Br J Pharmacol 1990; 100:879-885.

14. Kaumann AJ, Sanders L, Brown AM, Murray KJ, Brown MJ. A 5-HT$_{4\text{-like}}$ receptor in human right atrium. Naunyn-Schmiedeberg's Arch Pharmacol 1991; 344:150-159.

15. Ouadid H, Seguin J, Dumuis A, Bockaert J, Nargeot J. Serotonin increases calcium current in human atrial myocytes via the newly described 5-hydroxytryptamine$_4$ receptors. Mol Pharmacol 1992; 41:346-351.

16. Sanders L, Kaumann AJ. A 5-HT$_{4\text{-like}}$ receptor in human left atrium. Naunyn-Schmiedeberg's Arch Pharmacol 1992; 345:382-386.

17. Kaumann AJ, Sanders L. 5-Hydroxytryptamine causes rate-dependent arrhythmias through 5-HT$_4$ receptors in human atrium: facilitation by chronic β-adrenoceptor blockade. Naunyn-Schmiedeberg's Arch Pharmacol 1994; 349:331-337.

18. Kaumann AJ. Blockade of human atrial 5-HT$_4$ receptors by GR 113808. Br J Pharmacol 1993; 110:1172-1174.

19. Kaumann AJ, Gaster LM, King FD, Brown AM. Blockade of human atrial 5-HT$_4$ receptors by SB 207710, a selective and high affinity 5-HT$_4$ receptor antagonist. Naunyn-Schmiedeberg's Arch Pharmacol 1994; 349:546-548.

20. Kaumann AJ. Do human atrial 5-HT$_4$ receptors mediate arrhythmias? Trends Pharmacol Sci 1994; 15:451-455.

21. Sanders L, Lynham JA, Bond B, delMonte F, Harding SE, Kaumann AJ. Sensitization of human atrial 5-HT$_4$ receptors by chronic β blocker treatment. Circulation 1995; 92:2526-2639.

22. Kaumann AJ, Lynham JA, Brown AM. Comparison of the densities of 5-HT$_4$ receptors, β_1- and β_2-adrenoceptors in human atrium: functional implications. Naunyn-Schmiedeberg's Arch Pharmacol 1996; 353:592-595.

23. Kaumann AJ, Murray KJ, Brown AM, Frampton JE, Sanders L, Brown MJ. Heart 5-HT receptors. A novel 5-HT receptor in human atrium. In: Paoletti R, Vanhoutte P, Brunello N, Maggi FM, eds. Serotonin: From Cell Biology to Pharmacology & Therapeutics. Kluwers, Dordrecht, Boston, London 1989; 347-354.

24. Mohr B, Bom AH, Kaumann AJ, Thämer V. Reflex inhibition of the efferent renal sympathetic activity by 5-hydroxytryptamine and nicotine elicited by different epicardial receptors. Pflügers Arch 1987; 409:145-151.

25. Kaumann AJ, Brown AM. Allosteric modulation of arterial 5-HT$_2$ receptors. Saxena PR, Wallis DI, Wouters W, Bevan P, eds. Cardiovascular pharmacology of 5-hydroxytryptamine. Boston, London: Kluwers, Dordrecht, 1990:127-142.

26. Chester AH, Martin GR, Bodelsson M, Arneklo-Nobin B, Tadjkarimi S, Tornebrandt K, Yacoub M. 5-Hydroxytryptamine receptor profile in healthy and diseased human epicardial coronary arteries. Cardiovasc Res 1990; 24:932-937.

27. Kaumann AJ, Parsons AA, Brown AM. Human arterial constrictor 5-HT receptors. Cardiovasc Res 1993; 27:2094-2103.

28. Kaumann AJ, Frenken M, Posival H, Brown AM. Variable participation of 5-HT$_{1-like}$ receptors and 5-HT$_2$ receptors in serotonin-induced contraction of human isolated coronary arteries. Circulation 1994; 90:1141-1153.

29. Kaumann AJ, Murray KJ, Brown AM, Sanders L, Brown MJ. A receptor for 5-HT in human atrium. Br J Pharmacol 1989; 98:664P.

30. Dumuis A, Bouhelal R, Sebben M, Cory R, Bockaert J. A nonclassical 5-hydroxytryptamine receptor positively coupled with adenylate cyclase in the central nervous system. Mol Pharmacol 1988; 34:880-887.

31. Bockaert J, Sebben M, Dumuis A. Pharmacological characterization of 5-hydroxytryptamine$_4$ (5-HT$_4$) receptors coupled to adenylate cyclase in adult guinea pig hippocampal membranes: Effect of substituted benzamide derivatives. Mol Pharmacol 1990; 37:408-411.

32. Ansanay H, Sebben M, Bockaert J, Dumuis A. Characterization of homologous 5-hydroxytryptamine$_4$ receptor desensitization in colliculi neurones. Mol Pharmacol 1992; 42:808-816.

33. Dumuis A, Sebben M, Bockaert J. The gastrointestinal prokinetic benzamide derivatives are agonists at the non-classical 5-HT receptor (5-HT$_4$) positively coupled to adenylate cyclase in neurones. Naunyn-Schmiedeberg's Arch Pharmacol 1989; 340:403-410.

34. Bockaert J, Fozard J, Dumuis A, Clarke DE. The 5-HT$_4$ receptor: a place in the sun. Trends Pharmacol Sci 1992; 13:141-145.

35. Jahnel U, Rupp J, Ertl R, Nawrath H. Positive inotropic responses to 5-HT in human atrial but not in ventricular heart muscle. Naunyn-Schmiedeberg's Arch Pharmacol 1992; 346:482-485.

36. Jahnel U, Nawrath H, Rupp J, Ochi R. L-type calcium channel activity in human atrial myocytes as influenced by 5-HT. Naunyn-Schmiedeberg's Arch Pharmacol 1993; 348:396-402.

37. Turconi M, Schiantarelli P, Borsini F, Rizzi CA, Ladinsky H, Donetti A. Azabicycloalkyl benzimidazolones: Interaction with serotonergic 5-HT₃ and 5-HT₄ receptors and potential therapeutic implications. Drugs of the Future 1991; 19(11):1011-1026.

38. Schoemaker RG, DU XY, Bax WA, Bos E, Saxena PR. 5-Hydroxytryptamine stimulates human isolated atrium but not ventricle. Eur J Pharmacol 1993; 230:103-105.

39. Zerkowski H-R, Broede A, Kunde K, Hillemann S, Schäfer E, Vogelsang M, Michel MC, Brodde O-E. Comparison of the positive inotropic effects of serotonin, histamine, angiotensin II, endothelin and isoprenaline in the isolated human right atrium. Naunyn-Schmiedeberg's Arch Pharmacol 1993; 347-352.

40. Kaumann AJ, Brown AM, Raval P. Putative 5-HT₄₋ₗᵢₖₑ receptors in piglet left atrium. Br J Pharmacol 1991; 101:98 P.

41. Lorrain J, Grosset A, O'Connor E. 5-HT₄ receptors, present in piglet atria and sensitive to SDZ 205-557, are absent in papillary muscle. Eur J Pharmacol 1992; 229:105-108.

42. Hall JA, Kaumann AK, Brown MJ. Selective β-adrenoceptor blockade enhances positive inotropic responses to endogenous catecholamines mediated through β₂-adrenoceptors in human atrial myocardium. Circ Res 1990; 66:1610-1023.

43. Kaumann AJ, Hall JA, Murray KJ, Wells FC, Brown MJ. A comparison of the effects of adrenaline and noradrenaline on human heart: the role of adenylate cyclase and contractile force. Eur Heart J 1989; 10(Suppl B):29-37.

44. Medhurst AD, Kaumann AJ. Characterisation of the 5-HT₄ receptor mediating tachycardia in isolated piglet right atrium. Br J Pharmacol 1993; 110:1023-1030.

45. Craig DA, Clarke DE. Pharmacological characterization of a neuronal receptor for 5-hydroxytryptamine in guinea pig ileum with properties similar to the 5-hydroxytryptamine₄ receptor. J Pharmacol Exp Ther 1990; 252:1378-1386.

46. Elswood CJ, Bunce KT, Humphrey PPA. Identification of putative 5-HT₄ receptors in guinea-pig ascending colon. Eur J Pharmacol 1991; 196:149-155.

47. Baxter GS, Craig DA, Clarke DE. 5-Hydroxytryptamine₄ receptors mediate relaxation of rat oesophageal tunica muscularis mucosae. Naunyn-Schmiedeberg's Arch Pharmacol 1991; 343:439-446.

48. Ford APD, Baxter GS, Eglen RM, Clarke DE. 5-Hydroxytryptamine stimulates cyclic AMP formation in the tunica muscularis mucosae of the rat oesophagus via 5-HT₄ receptors. Eur J Pharmacol 1992; 211:117-120.

49. McLean PG, Coupar IM. 5-HT₄ receptor antagonist affinities of SB 207710, SB 205008 and SB 203186 in human colon, rat oesophagus and guinea-pig ileum peristaltic reflex. Naunyn-Schmiedeberg's Arch Pharmacol 1995; 352:132-140.

50. Gerald C, Adham N, Kao H-T, Olsen MA, Laz TM, Schechter LE, Bard JA, Vaysee Pj-J, Hartig PR, Branchek TA, Weinshank RL. The 5-HT$_4$ receptor: molecular cloning and pharmacological characterization of two splice variants. EMBO J 1995; 14:2806-2815.

51. Blondel O, Vandecasteele G, Gastineau M, Leclerc S, Dahmoune Y, Langlois M, Fischmeister R. Molecular and functional characterization of a 5-HT$_4$ receptor cloned from human atrium. FEBS Letters 1997; 412:465-474.

52. Claeysen S, Sebben M, Journot L, Bockaert J, Dumuis A. Cloning, expression and pharmacology of the mouse 5-HT$_{4L}$ receptor. FEBS Letters 1996; 398:19-25.

53. Ullmer C, Schmuck K, Kalkman HO, Lübbert H. Expression of serotonin receptor mRNAs in blood vessels. FEBS Letters 1995; 370:215-221.

54. Ouadid H, Albat B, Nargeot J. Calcium currents in diseased cardiac cells. J Cardiovasc Pharmacol 1995; 25:282-291.

55. Beukelmann DJ, Erdmann E. Ca^{2+} currents and intracellular [Ca^{2+}] transients in single ventricular myocytes isolated from terminally failing human myocardium. Basic Res Cardiol 1992; 87(Suppl 1):235-243.

56. Sanders L, Lynham JA, Kaumann AJ. Chronic β$_1$-adrenoceptor blockade sensitises the H$_1$ and H$_2$ receptor systems in human atrium: role of cyclic nucleotides. Naunyn-Schmiedeberg's Arch Pharmacol 1996; 353:661-670.

57. Molenaar P, Sarsero D, Arch JRS, Kelly J, Henson SM, Kaumann AJ. Effects of (-)-RO363 at human atrial β-adrenoceptor subtypes, the human cloned β$_3$-adrenoceptor and rodent intestinal β$_3$-adrenoceptors. Br J Pharmacol 1997; 120:165-176.

58. Kaumann AJ. (-)-CGP 12177-induced increase of human atrial contraction through a putative third β-adrenoceptor. Br J Pharmacol 1996; 117:93-98.

59. Kaumann AJ, Molenaar P. Modulation of human cardiac function through 4 β-adrenoceptor populations. Naunyn-Schmiedeberg's Arch Pharmacol 1997; 355:667-681.

60. Kaumann AJ, Lynham JA, Sanders L, Brown AM, Molenaar P. Contribution of differential efficacy to the pharmacology of human β$_1$- and β$_2$-adrenoceptors. Pharmacol Commun 1995; 6:215-222.

61. Hausdorff WP, Pitcher JA, Luttrell DK, Linder ME, Kurose H, Parsons SJ, Caron MG, Lefkowitz RJ. Tyrosine phosphorylation of G protein subunits by pp60^{c-src}. Proc Natl Acad Sci USA 1992; 89:5720-5724.

62. Gille E, Lemoine H, Ehle B, Kaumann AJ. The affinity of (-)-propranolol for β$_1$- and β$_2$-adrenoceptors of human heart. Differential antagonism of the positive inotropic effects and adenylate cyclase stimulation by (-)-noradrenaline and (-)-adrenaline. Naunyn-Schmiedeberg's Arch Pharmacol 1985; 331:60-70.

63. Kaumann AJ, Lemoine H. β₂-Adrenoceptor-mediated positive inotropic effect of adrenaline in human ventricular myocardium. Quantitative discrepancies with binding and adenylate cyclase stimulation. Naunyn-Schmiedeberg's Arch Pharmacol 1987; 335:403-411.

64. Green SA, Holt BD, Liggett SB. β₁- and β₂-adrenergic receptors display subtype-selective coupling to Gs. Mol Pharmacol 1992; 41:889-893.

65. Levy F-O, Zhu X, Kaumann AJ, Birnbaumer L. Efficacy of β₁-adrenergic receptors is lower than that of β₂-adrenergic receptors. Proc Natl Acad Sci USA 1993; 90:10798-10802.

66. Sole MJ, Shum A, VanLoon GR. Serotonin: metabolism in the normal and failing heart. Circ Res 1979; 45:629-634.

67. Cohen RA. Platelet-induced neurogenic coronary contractions due to accumulation of the false neurotransmitter 5-hydroxytryptamine. J Clin Invest 1985; 75:286-292.

68. Shah AM, Andries LJ, Meulemans AL, Brutsaert DL. Endocardium modulates myocardial inotropic responses to 5-hydroxytryptamine. Am J Physiol 1989; 257: H1790-H1797.

69. DiFrancesco D. Pacemaker mechanisms in cardiac tissues. Ann Rev Physiol 1993; 55:451-467.

70. Pino R, Cerbai E, Alajamo F, Porciatti F, Calamai G, Mugelli A. Effect of 5-hydroxytryptamine (5-HT) on the pacemaker current, i_f, in human atrial myocytes. Circulation 1996; 94: I-473 (abstract 2770).

71. Kannel WB, Abbott RD, Savage DD, McNamara PM. Epidemiological factors of atrial fibrillation: the Framingham study. N Engl J Med 1982; 306:1018-1022.

72. Bialy D, Lehmann MH, Schumacher DN, Steinman RT, Meissner MD. Hospitalization for arrhythmias in the United States: importance of atrial fibrillation. JACC 1992; 19:41.

73. Probst P, Goldschlager N, Selzer A. Left atrial size and atrial fibrillation in mitral stenosis: factors influencing this relationship. Circulation 1973; 48:1282-1287.

74. Henry WL, Morganroth J, Pearlman AS, Clark CE, Redwood DR, Itscoitz SB, Epstein SE. Relation between echocardiographically determined left atrial size and atrial fibrillation. Circulation 1976; 53:273-279.

75. Petersen P, Kastrup J, Brinch K, Godtfredsen J, Boysen G. Relation between left atrial dimension and duration of atrial fibrillation. Am J Cardiol 1987; 60:382-384.

76. Takahashi N, Imataka K. Seki A, Fujii J. Left atrial enlargement in patients with paroxysmal atrial fibrillation. Jap Heart J 1982; 23:677-683.

77. Keren G, Etzion T, Sherez J, Selcer AA, Megidish R, Miller HI, Laniado S. Atrial fibrillation and atrial enlargement in patients with mitral stenosis. Am Heart J 1988; 114:1146-1155.

78. Nunain SO, Debbas NMG, Camm AJ. Determinants of the course and prognosis of atrial fibrillation. In: Atrial arrhythmias. Current

Concepts and Management. Touboul P, Waldo AL, eds. Mosby Year Book, Boston 1990; 350-358.

79. Daniel WG, Nellessen U, Schröder Em Nonnast-Daniel B, Bednarski P, Nikutta P, Lichtlen PR. Left atrial spontaneous echo contrast in mitral valve disease: an indicator for an increased thromboembolic risk. JACC 1988; 11:1204-1211.

80. Halperin JL, Hart RG. Atrial fibrillation and stroke: new ideas, persisting dilemmas. Stroke 1988; 19:937-941.

81. Kulbertus HE. Thromboembolism in atrial fibrillation. In: Atrial arrhythmias. Current Concepts and Management. Touboul P, Waldo A, eds. Mosby Year Book, Boston 1990; 359-361.

82. Wolf PA, Dawler TR, Thomas HE, Kannel WB. Epidemiologic assessment of chronic atrial fibrillation and risk of stroke: the Framingham study. Neurology 1978; 28:973-294.

83. Petersen P, Godtfredsen J. Risk factors for stroke in chronic atrial fibrillation. Eur Heart J 1988; 9:291-294.

84. Lundin L, Norheim I, Landelius J, Oberg K, Theodorsson-Norheim E. Carcinoid heart disease: relationship of circulating vasoactive substances to ultrasound-detectable cardiac abnormalities. Circulation 1988; 77:264-269.

85. Pellikka PA, Tajik J, Khandeira BK, Seward JB, Callahan JA, Pitot HC, Kvols LK. Carcinoid heart disease. Clinical and echocardiographic spectrum in 74 patients. Circulation 1993; 87:1188-1196.

86. Olssen S, Edwards IR. Tachycardia during cisapride treatment. Br Med J 1992; 305:748-749.

87. Inman W, Kunota K. Tachycardia during cisapride treatment. Br med J 1992; 305:1019.

88. Harris P, Fritts HW, Cournand A. Some circulatory effects of 5-hydroxytryptamine in man. Circulation 1960; 21:1134-1139.

89. Hollander W, Michelson AL, Wilkins RW. Serotonin and antiserotonins. I. Their circulatory, respiratory and renal effects in man. Circulation 1957; 16:246-255.

90. Le Mesurier DH, Schwartz CJ, Whelan RF. Cardiovascular effects of intravenous infusions of 5-hydroxytryptamine in man. Br J Pharmacol 1959; 14:246-250.

91. Parks VJ, Sandison AG, Skinner SL, Whelan RF. The stimulation of respiration by 5-hydroxytryptamine in man. J Physiol 1960; 151:342-351.

92. Rhodes KF, Coleman J, Lattimer A. A component of 5-HT-evoked depolarisations of the rat isolated vagus is mediated by a putative $5-HT_4$ receptor. Naunyn-Schmiedeberg's Arch Pharmacol 1992; 346:496-503.

93. Villalon CM, DenBoer MO, Heiligers JAC, Saxena PR. Further characterization by the use of tryptamine and benzamide derivatives, of the putative $5-HT_4$ receptor mediating tachycardia in the pig. Br J Pharmacol 1991; 102:107-112.

94. Bateman DN. The action of cisapride on gastric emptying and the pharmacodynamics and pharmacokinetics of oral diazepam. Eur J Clin Pharmacol 1986; 30:205-208.

95. Kaumann AJ. 5-Hydroxytryptamine and the human heart. Fozard JR, Saxena PR, eds, Serotonin: Molecular Biology, Receptors and Functional Effects. Birkhauser Basel Switzerland 1991; 365-373.

96. Humphrey PPA, Bunce KT. Tachycardia during cisapride treatment. Br med J 1992; 305:1019-1920.

97. Saxena PR, Villalon CM. 5-Hydroxytryptamine: a chameleon in the heart. Trends Pharmacol Sci 1991; 12:223-227.

98. Curtis MJ, Pugsley MK, Walker MJA. Endogenous chemical mediators of ventricular arrhythmias in ischaemic heart disease. Cardiovasc Res 1993; 27:703-719.

99. Willet F, Curzen N, Adams J, Armitage M. Coronary vasospasm induced by subcutaneous sumatriptan. Br Med J 1992; 304:1215.

100. Ottervanger JP, Paalman HJA, Boxma GL, Stricker BHCh. Transmural myocardial infarction with sumatriptan. Lancet 1993; 341:861-862.

101. Kaumann AJ. Adrenergic receptors in heart muscle: relations among factors influencing the sensitivity of the cat papillary muscle to catecholamines. J Pharmacol Exp Ther 1970; 173:383-398.

CHAPTER 7

5-HT$_4$ Receptors in Gastrointestinal Tract

Sharath S. Hegde

Introduction

The precise control of gastrointestinal motor and secretory activity is critically dependent on neurotransmitters and neuromodulators contained within the enteric nervous system and endocrine cells of the gut. Enteric 5-hydroxytryptamine (5-HT) which accounts for a large proportion of the total body 5-HT,[1] is a key transmitter which exerts profound effects on various aspects of gastrointestinal function.[2,3] 5-HT, which can be released from either enterochromaffin cells[1] or enteric neurons,[4] can elicit a plethora of responses in the gut, including modulation of neurotransmitter release, direct smooth muscle contractile/relaxant effects, stimulation of enterocyte secretory mechanisms, modulation of peristaltic reflex and activation of sensory nerves.[2,3] Given the potential pathophysiological role of endogenous 5-HT in certain gastrointestinal disorders and the multiplicity of 5-HT receptors (at least seven subtypes of 5-HT receptors have been identified thus far), it is not surprising that research into the pharmacological characterization of 5-HT receptors in the gut has sparked colossal interest. The 5-HT$_4$ receptor in the gut,[5,6] in particular, has attracted enormous attention primarily because the presence of peripheral 5-HT$_4$ receptors was first demonstrated in the ileum;[7] this discovery has led to a better understanding of the mechanism of action of existing prokinetic drugs such as cisapride and may set the stage for novel gastrointestinal therapies.

5-HT$_4$ Receptors in the Brain and Periphery, edited by Richard M. Eglen.
© 1998 Springer-Verlag and R.G. Landes Company.

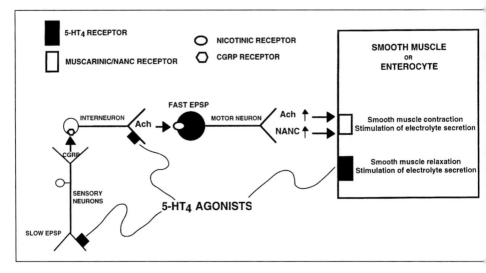

Fig. 7.1. Schematic diagram showing the location and effects of 5-HT₄ receptors in the enteric nervous system. 5-HT₄ receptor agonists can facilitate nicotinic ganglionic transmission (fast EPSP) in myenteric or submucosal interneurons that results in augmented release of Ach or NANC transmitters from motor neurons innervating smooth muscle or enterocyte, leading to smooth muscle contraction or stimulation of electrolyte secretion, respectively. In certain tissues, 5-HT₄ receptor agonists act directly on the smooth muscle or enterocyte to cause smooth muscle relaxation or stimulation of electrolyte secretion, respectively. 5-HT₄ receptor activation can also sensitize CGRP-containing sensory neurons by facilitating slow depolarization (slow EPSP). Ach = acetylcholine, CGRP = Calcitonin gene related peptide, NANC = nonadrenergic noncholinergic, EPSP = excitatory postsynaptic potential.

Localization and Coupling of 5-HT₄ Receptors in the Gastrointestinal Tract

High affinity specific 5-HT₄ binding sites, using [³H]-GR-113808 as the radioligand, have been detected throughout the intestine of the guinea pig with the regional rank order of receptor density being: duodenum > jejunum > ileum > colon > ileum.[8] Reverse transcriptase-polymerase chain reaction (RT-PCR) studies have also detected the receptor transcript in several gastrointestinal tissues of the rat.[9] Functional studies (see below) suggest that 5-HT₄ receptors are localized on myenteric and submucosal neurons, smooth muscle and on mucosal enterocytes. As in the central nervous system, these receptors are positively coupled to adenylate cyclase in gastrointestinal tissues such as the esophagus[10] and colon.[11] It has been proposed that, similar to observations in neurons of the central nervous system,[12] 5-HT₄ receptor activation in myenteric neurons causes stimulation of cAMP-dependent protein kinase A and, consequently, closure of potassium channels. The resulting depolarization may underlie the neuronal excitatory effects of $5\text{-}HT_4$ receptor agonists in the gut. Also, $5\text{-}HT_4$ receptor-mediated relaxation of certain gastrointestinal tissues may be caused by cAMP-induced reduction of intracellular Ca^{2+} levels.

A characteristic feature of the $5\text{-}HT_4$ receptor is its propensity to desensitize upon continuous stimulation.[5,6] This property is also shared by $5\text{-}HT_4$ receptors mediating facilitation of neurotransmitter release in the guinea pig ileum[13] and relaxation of rat esophagus.[14] In the latter tissue, the desensitization process has been studied extensively and shown to be biphasic, consisting of an initial rapid phase (~2 min) during which ~ 50% desensitization occurs and a slower second phase (~2 hours) during which desensitization is complete.[14] The limiting step appears to be at the level of the receptor itself or in its coupling to adenylate cyclase in as much as the kinetics of desensitization for cAMP stimulation and relaxation are similar. Since total receptor density is unchanged, uncoupling of the $5\text{-}HT_4$ receptor to its effector is the most likely mechanism of desensitization involved. The desensitization process in the rat esophagus is not simply an in vitro phenomenon since it can be demonstrated after chronic infusion of 5-HT to rats.[15] This finding may be clinically relevant since selective $5\text{-}HT_4$ receptor agonists are in development as potential prokinetic agents.

Effects on Smooth Muscle Tone

Although the vast majority of $5\text{-}HT_4$ receptor-mediated effects on smooth muscle tone appear to be stimulatory, inhibitory responses have been observed in a few instances. In general, the stimulatory responses are mediated indirectly via facilitation of neuronal transmission whereas the inhibitory responses are caused by direct smooth muscle relaxation.

Stimulatory Effects

In the guinea pig ileum, the neuronal receptor mediating atropine-sensitive contractions was originally classified as the M receptor[16] using the nomenclature proposed by Gaddum and Picarelli.[17] However, in 1990, soon after the demonstration of a novel $5\text{-}HT_4$ receptor in mouse colliculi neurons, Craig and Clarke[7] showed that 5-HT-induced augmentation of cholinergically mediated contractions in the electrically stimulated longitudinal smooth muscle myenteric plexus preparation of guinea pig ileum was mediated via a receptor which bore close pharmacological resemblance to the 5-HT_4. Similar findings were also reported, subsequently, for the stimulated circular muscle of the guinea pig ileum.[18] Studies using the unstimulated guinea pig ileum showed that the high potency phase of the contractile response to 5-HT could also equated, pharmacologically, with the 5-HT_4 receptor.[19] Likewise, 5-HT-induced stimulation of phasic and tonic activity of the unstimulated human ileum also appears to be 5-HT_4 receptor-mediated.[20]

Direct evidence for 5-HT_4 receptor-mediated facilitation of cholinergic transmission was provided by Kilbinger et al,[21,22] who showed that both basal and electrically-evoked outflow of [³H]-acetylcholine from the guinea pig longitudinal myenteric plexus preparation was enhanced by 5-HT_4 receptor agonists, and these effects could be antagonized by either tropisetron (pA_2 estimate ~ 6.6) or DAU 6285 (pA_2 estimate ~ 7.1). Since the release of [³H]-acetylcholine was markedly inhibited by (+)-tubocurarine, it was suggested that 5-HT_4 receptors are located on cholinergic interneurones where they facilitate nicotinic ganglionic transmission—a notion that concurs with electrophysiological studies which have shown that 5-HT_4 receptor stimulation facilitates ganglionic transmission (fast EPSP) in myenteric neurons (type S) in guinea pig ileum.[23,24] King and Sanger[25] have additionally proposed that 5-HT_4 receptor activation facilitates release of non-cholinergic transmitters in the guinea pig ileum. It is possible that neurokinin B may be at least one of the potential noncholinergic transmitters as it has been shown that NK_3 receptor desensitization suppresses the atropine-resistant component of the contractile response evoked by 5-methoxytryptamine (5-MeOT) in the guinea pig ileum.[26] 5-HT_4 receptor activation has also been shown to enhance slow depolarization (slow EPSP) in sensory (AH-type) neurons[23] and in this manner increase their responsiveness to other stimuli.

5-HT_4 receptors also mediate contraction of the proximal[27-30] and distal colon[31,32] of the guinea pig via facilitation of cholinergic and noncholinergic transmission. The guinea pig distal colon is unique in that 5-HT is exceedingly potent ($pEC_{50} = 9.2$), perhaps reflecting the high reserve of 5-HT_4 receptors in this tissue. The effects of 5-HT on the proximal colon have been confirmed in vivo in the anesthetized guinea pig in which 5-HT-induced contraction of the circular muscle is potently antagonized by SB 204070, a selective 5-HT$_4$ receptor antagonist.[33] The dog[34] and human colon,[35] surprisingly, lack contractile 5-HT$_4$ receptors, but may possess relaxant 5-HT$_4$ receptors (see below).

Besides modulation of cholinergic and noncholinergic transmission in the guinea pig intestine, 5-HT$_4$ receptors can also modulate 5-HT release from enterochromaffin cells. In the vascularly perfused guinea pig small intestine, 5-MeOT, BIMU 8 and cisapride were shown to produce concentration-dependent inhibition of 5-HT release.[36] Interestingly, high concentrations of tropisetron enhance 5-HT release, suggesting that release of endogenous 5-HT, from enterochromaffin cells, is tonically inhibited by 5-HT$_4$ receptors. Studies in the isolated rat ileum, however, have shown a stimulatory effect of 5-MeOT on 5-HT release although the involvement of 5-HT$_4$ receptors in this response was not unequivocally established.[37]

It has long been known that benzamides, such as cisapride, are potent gastroprokinetic agents in several species, including man.[38] The mechanism underlying their gastroprokinetic effects has been the subject of considerable controversy. Following the exclusion of a dopaminergic mechanism, a 5-HT receptor involvement was proposed. However, as these compounds possess 5-HT$_3$ antagonistic and 5-HT$_4$ agonistic properties, the precise underlying mechanisms (5-HT$_3$ or 5-HT$_4$) have, until recently, been obscure. In dogs, early indications for the involvement of 5-HT$_4$ receptors in gastroprokinesis were provided by Gullikson et al,[39] who showed that SC 49518, a benzamide compound possessing selectivity for the 5-HT$_4$ receptor, could reverse α_2-adrenoceptor-mediated delay in gastric emptying. Rizzi et al[40] showed that the gastroprokinetic effects of BIMU 1 in dogs could be antagonized by DAU 6285, a marginally selective 5-HT$_4$ receptor antagonist. Furthermore, it was recently reported that the contractile response to 5-HT in the Heidenhan pouch of the conscious dog in vivo[41] and the dog stomach antrum in vitro[38] could be antagonized by the selective 5-HT$_4$-receptor antagonist SB 204070. The gastroprokinetic effects of benzamides and benzimidazolones in rats has previously been attributed to 5-HT$_3$ receptor antagonism[42] but this contention has, however, been questioned by recent findings. LY 277359, a potent and selective 5-HT$_3$ receptor antagonist, is devoid of gastroprokinetic activity in rats.[43] In conscious rats, Hegde et al[44] showed that SC 49518 produces potent stimulation of gastric emptying at doses that are 150-fold lower than that required to inhibit 5-HT$_3$ receptor-mediated von Bezold-Jarisch reflex. Furthermore, the gastroprokinetic effects SC 49518 are antagonized by the selective 5-HT$_4$ receptor antagonists, RS 23597 and SB 204070. Additional supportive evidence for the involvement of 5-HT$_4$ receptors was provided by Gale et al[45] who showed that the gastroprokinetic effect of metocl'opramide in rats could be completely inhibited by GR 125487, a selective 5-HT$_4$ receptor antagonist. It is also pertinent to note that contractile 5-HT$_4$ receptors have recently been identified in the rat stomach fundus.[46] 5-HT$_4$ receptors are also located in the guinea pig stomach where they facilitate electrically-evoked cholinergically-mediated contractions.[47,48] Likewise, in humans, cisparide, 5-HT and 5-MeOT have been shown to enhance electrically-induced contractions of the fundus, corpus and antrum via 5-HT$_4$ receptors.[49] The sheep is the only species in which inhibitory 5-HT$_4$ receptors are present on the forestomach[50] (see below).

Inhibitory Effects

5-HT-induced relaxation of the precontracted rat esophagus, that was originally attributed to 5-HT$_3$ receptor activation, was subsequently shown to result from 5-HT$_4$ receptor stimulation.[51,52] Due to the robustness of this preparation, it is widely employed as a reliable screening assay for 5-HT$_4$ receptor ligands. 5-HT also induces relaxation of the rat isolated ileum via a receptor whose pharmacological profile indicates the involvement of 5-HT$_4$ receptors.[53] Interestingly, studies in both the isolated esophagus and ileum of rat have demonstrated the existence of an endogenous 5-HT$_4$ receptor-mediated relaxant tone that can be abolished by pretreatment of the animals with p-chlorophenylalanine.[54,55]

5-HT has been shown to inhibit spontaneous contractions of the circular muscle in human isolated colon via, predominantly, a non-neural mechanism.[35] The rank order of potency of several indole and benzamide agonists together with the low antagonistic affinity of tropisetron ($pA_2 = 6.0$) indicates the involvement of 5-HT$_4$ receptors in the inhibitory effects of 5-HT.[35] However, subsequent studies showed that the highly selective 5-HT$_4$ receptor antagonist GR 113808 behaves anomalously in the human colon.[56,57] At low concentrations (3 nM), GR 113808 displaced the 5-HT concentration-effect curve in a parallel manner (pA_2 estimate = 8.9), whereas higher concentrations (10-100 nM) produced no further displacement of the 5-HT curve. Furthermore, another 5-HT$_4$ receptor antagonist SDZ

205-557 (0.3-10 μM) failed to antagonize the relaxant responses to 5-HT. These discrepant findings suggest that the relaxant responses to 5-HT may be mediated by multiple receptors, one of which resembles the $5\text{-}HT_4$ receptor.

Relaxant $5\text{-}HT_4$ receptors may also be present in the dog colon in as much as the contractions to 5-HT are potentiated by the $5\text{-}HT_4$ antagonist SB 204070 in this tissue.[34] Also, $5\text{-}HT_4$ receptors mediating forestomach hypomotility have been reported and this is the only case in which the inhibitory effects of the receptor are neuronally mediated.[50]

Peristaltic Reflex

Intestinal propulsion is critically dependent on the operation of a local enteric reflex termed the 'peristaltic reflex'.[58] This reflex, which is evoked in response to gut distension or mucosal stimulation, leads to ascending contraction and descending relaxation of circular muscle. $5\text{-}HT_4$ receptor agonism, either on sensory nerves or on interneuronal synapse in the myenteric plexus, can potentially modulate this reflex.

In the isolated ileum of the guinea pig, 5-HT, 5-MeOT and numerous prokinetic benzamides can facilitate the peristaltic reflex, induced by distension, with a rank order of potency that correlates with their relative efficacy for $5\text{-}HT_4$ receptors.[59-61] Furthermore, the facilitatory effects of 5-HT can be antagonized by both SDZ 205-557 and high concentrations of tropisetron, implicating the involvement of $5\text{-}HT_4$ receptors. $5\text{-}HT_4$ receptor-mediated facilitation of the peristaltic reflex by 5-HT and 5-MeOT has also been demonstrated in the marmoset ileum.[62] In all the above studies, $5\text{-}HT_4$ receptor antagonists alone failed to modify the reflex, implying that normal operation of the reflex is not critically dependent on $5\text{-}HT_4$ receptor activation, a conjecture that is corroborated by a recent study that showed no effect of $5\text{-}HT_4$ receptor antagonists on the ascending excitatory and descending inhibitory reflex that is elicited in response to luminal distension of the guinea-pig intestine.[63] However, a recent study in which the peristaltic reflex was assessed by measuring the propulsion of solid pellets in isolated segments of the guinea pig distal colon, showed that while neither selective $5\text{-}HT_4$ or $5\text{-}HT_3$ receptor antagonists affected the reflex, the combined blockade of the two receptors produced significant inhibition of the reflex, suggesting redundancy in the 5-HT pathway.[64]

Table 7.1. 5-HT$_4$ receptor mediated effects in gastrointestinal tissues of different species

Organ	Species	Experimental Conditions	Effect	Refs.
Esophagus	Rat	Isolated tunica muscularis mucosae	Relaxation	51, 52
Stomach	Man	Isolated stomach antrum, fundus	Potentiation of electrically-evoked contractions	49
	Dog	Isolated stomach antrum	Potentiation of electrically-evoked contractions	38
		Conscious dog	Stimulation of gastric emptying, contraction of gastric Heidenhein pouch	39-41
	Rat	Isolated stomach fundus	Potentiation of spontaneous and electrically-evoked contractions	46
		Conscious rat	Stimulation of gastric emptying	44, 45
	Guinea pig	Isolated circular muscle/strips of stomach	Potentiation of electrically-evoked contractions	47, 48
	Sheep	Conscious sheep	Inhibition of forestomach motility	50
Jejunum	Man	Isolated jejunal mucosa	Stimulation of short-circuit current	76
		Isolated jejunal segments	Facilitation of mucosal stimuli-induced peristaltic reflex	66
	Pig	Isolated muscle-stripped jejunum	Stimulation of short-circuit current	74
Ileum	Man	Isolated ileal mucosa	Stimulation of short-circuit current	75
	Marmoset	Isolted whole ileum	Facilitation of peristaltic reflex	62

	Species	Preparation	Effect	Ref.
	Rat	Isolated ileum	Relaxation	53
		Anesthetized rat	Stimulation of ileal secretion	73
	Guinea pig	Isolated longitudinal myenteric-plexus or circular muscle of ileum	Potentiation of electrically-evoked contractions	7, 18
		Isolated whole ileum	Contraction	19
		Isolated whole ileum	Facilitation of peristaltic reflex	59-61
		Isolated ileal mucosa	Stimulation of short-circuit current	68-70
		Isolated myenteric plexus	Enhancement of acetylcholine release	21, 22
		Isolated longitudinal myenteric plexus	Facilitation of fast-EPSP	24
		Myenteric neurons	Potentiation of slow-EPSP and fast EPSP	23
Colon	Man	Isolated circular muscle of colon	Inhibition of spontaneous contractions	35, 56, 57
		Isolated colonic mucosa	Stimulation of short-circuit current	77
	Rat	Isolated colonic mucosa	Stimulation of short-circuit current	71
		Isolated colonic segments	Facilitation of mucosal stimuli-induced peristaltic reflex	65
		Anesthetized rat	Stimulation of colonic secretion	72
	Guinea pig	Isolated proximal and distal colon	Contraction	27-32
		Isolated colonic segments	Facilitation of mucosal stimuli-induced peristaltic reflex	66
		Anesthetized guinea-pig	Contraction of colon	33
Other effects in gut	Dog	Conscious dog	Copper sulfate-induced emesis	80
	Ferret	Conscious ferret	Copper sulfate and zacopride-induced emesis	81
	Mice	Conscious mice	Stimulation of diarrhea and fecal output	98, 99

In contrast to the aforementioned findings, investigations in which the peristaltic reflex to selective mucosal stimulation have been studied have revealed a more critical role of $5\text{-}HT_4$ receptors in the operation of the reflex. Experiments in the rat colon,[65] guinea pig colon[66] and human jejunum[66] have shown that application of mucosal stimuli (2-10 brush strokes) evokes ascending contraction and descending relaxation of circular muscle that is accompanied by release of CGRP and 5-HT into the medium. All these effects are selectively inhibited by the $5\text{-}HT_4$ receptor antagonist SDZ 205-557. In these studies, muscle stretch evoked CGRP but not 5-HT release; the accompanying ascending contraction and descending relaxation were not affected by SDZ 205-557. These data suggest that $5\text{-}HT_4$ receptor activation by endogenous 5-HT plays an integral part in initiating the peristaltic reflex, evoked by mucosal stimulation but not muscle stretch, by stimulating CGRP containing sensory nerves which subsequently activates the motor limb of the reflex. The differential role of $5\text{-}HT_4$ receptors in peristalsis induced by mucosal and stretch stimuli may be related to the distinct sensory pathways which are activated in response to the two stimuli—mucosal stimuli activate intrinsic sensory neurons whose cell bodies are located within the wall of the intestine whereas stretch stimuli activate extrinsic sensory neurons whose cell bodies are localized in the dorsal root ganglia.[67]

Secretion

5-HT is a potent intestinal secretagogue in several species including man. Measurement of short-circuit current (SCC) response in isolated mucosal preparations of the gut, which serves as an index of net mucosal electrolyte secretion, has been commonly used to study the secretory effect of 5-HT in vitro.

In the guinea pig isolated ileal mucosal preparation, 5-HT stimulates an increase in SCC via both tetrodotoxin (TTX)-sensitive and insensitive mechanisms.[68-70] The concentration-effect curve for the TTX-sensitive component is biphasic with the high potency phase being mediated by $5\text{-}HT_4$ receptors in as much as it can be mimicked by 5-MeOT and SC 53116 and antagonized by GR 113808 (pA_2 estimate = 9.6).[68] In contrast, the TTX-insensitive responses to 5-HT are resistant to $5\text{-}HT_1$, $5\text{-}HT_2$, $5\text{-}HT_3$ and $5\text{-}HT_4$ receptor antagonists.[69,70] However, paradoxically, the TTX-insensitive responses to 5-MeOT and R,S-zacopride are antagonized by tropisetron (pA_2 estimate ~ 6) consistent with the involvement of $5\text{-}HT_4$ receptors.[69]

These findings suggest that although neuronally and non-neuronally mediated secretory responses to 5-HT$_4$ receptor stimulation can be demonstrated in the guinea pig ileal mucosa, it is only the former mechanism that dominates the response to the endogenous agonist 5-HT. In rat colonic mucosa, the secretory responses to 5-HT and 5-MeOT are largely mediated via extraneuronal 5-HT$_4$ receptors as they are resistant to TTX and antagonized by tropisetron (pA$_2$ = 6) and GR 113808 (pA$_2$ = 9.8).[71] In the pentobarbital anesthetized rat, intravenous 5-HT and 5-MeOT evoke increases in transcolonic potential difference that was significantly inhibited, albeit marginally, by the selective 5-HT$_4$ receptor antagonist SB 204070, thereby confirming the in vitro findings.[72] In addition, 5-HT$_4$ receptors also appear to mediate, at least in part, the secretory response to intraluminally administered 5-HT in the anesthetized rat.[73] A recent study has suggested that the secretory responses to 5-HT and 5-MeOT in the isolated jejunum of the pig are mediated by 5-HT$_4$ receptors.[74] However, a surprising finding of this study was that, although high concentrations of tropisetron (10 μM) were required to antagonize the effects of 5-HT, relatively low concentrations (10 and 100 nM) were needed to block the effects of 5-MeOT. An equally unusual observation of the study was that tropisetron behaved as a weak partial agonist. A non-neuronally located 5-HT$_4$ receptor has also been implicated in the secretory effects of 5-HT in the ileal,[75] jejunal[76] and colonic[77] mucosa of man.

Emesis

5-HT$_4$ receptors are present on vagal afferents, activation of which mediates a long lasting depolarization.[78,79] As activation of abdominal vagal afferents can stimulate the 'vomiting center' in the central nervous system, it has been suggested that 5-HT$_4$ receptors may play a role in the emetic response. In support of this suggestion, 5-HT$_4$ receptors have been proposed to mediate vomiting in dogs following oral administration of copper sulfate, since the effect was blocked by high concentrations of tropisetron.[80] Bhandari and Andrews[81] have shown that high concentrations of tropisetron also block vomiting induced by copper sulfate and the nonselective 5-HT$_4$ receptor agonist, zacopride in ferrets. In contrast, Qin et al[82] have suggested that 5-HT$_4$ receptors play no role in vomiting induced by erythromycin. It has been suggested that 5-HT$_4$ receptor antagonists may play a role in modulating cytotoxic drug-induced nausea and vomiting.[83] Drugs of this kind induce emesis by release of 5-HT from enterochromaffin cells, which activates 5-HT$_3$ receptors. However, selective 5-HT$_3$ receptor antagonists reduce the vomiting in the clinic by only 50-60%. Although highly speculative, it is possible that the concurrent use of a selective 5-HT$_4$ receptor antagonist may augment the maximal therapeutic efficacy.

Therapeutic Value of Selective 5-HT$_4$ Receptor Agonists and Antagonists

Gastrointestinal Prokinetic Agents

Gastrointestinal prokinetic benzamides, possessing 5-HT$_4$ receptor agonist properties, have been employed in the treatment of gastro-esophageal reflux (heartburn), gastroparesis, functional dyspepsia and constipation.[38,83] This class of drugs promote coordinated gut wall contractions and augment propulsive activity in the caudal direction.

Metoclopramide, which belongs to the first generation class of compounds, is an antagonist at dopamine and 5-HT$_3$ receptors, besides being a weak partial 5-HT$_4$ receptor agonist.[38] Cisapride, renzapride and zacopride, compounds belonging to the second generation class of compounds, possess little or no affinity for dopamine receptors yet retain 5-HT$_4$ receptor agonistic and 5-HT$_3$ receptor antagonistic properties.[38] The benzimidazolones, BIMU 1 and BIMU 8,[84] and the benzothiozole, VB 201B7,[85] also possess similar pharmacological properties. The gastroprokinetic properties of these compounds have been ascribed, largely, to agonism of 5-HT$_4$ receptors. Cisapride, the most widely used gastroprokinetic, has been demonstrated to be effective in the treatment of gastro-esophageal reflux disease, functional dyspepsia and gastric stasis of various etiologies.[86] Although this drug enhances lower esophageal sphincter tone, a property that underlies its efficacy in reflux esophagitis, definitive evidence for a 5-HT$_4$ receptor involvement in this process is lacking. Cisapride has also been shown to facilitate colonic peristalsis in acute and chronic intestinal pseudo-obstruction and increase stool frequency in patients with chronic constipation.[86] Paradoxically, 5-HT$_4$ receptors in the human colonic smooth muscle are inhibitory. Thus, although a non-5-HT$_4$ receptor mechanism cannot be discounted, it is equally plausible that agonism of $5\text{-}HT_4$ receptors at an additional site(s), perhaps central, may account for the stimulatory effects of cisapride on lower gut motility in man.

Several third generation $5\text{-}HT_4$ receptor agonists, including the benzamide SC 49518,[39] the benzoate ML 10302,[87] the aryl ketone RS-67506[88] and the indole carbazimidamide SDZ HTF919[89] have been recently developed. Unlike the second generation compounds, these drugs have no appreciable interaction with $5\text{-}HT_3$ receptors and thus represent the first truly selective, albeit partial, $5\text{-}HT_4$ receptor agonists. SDZ HTF919, the drug which is most advanced in clinical development, has been shown to decrease colonic transit time and produce loose stools, consistent with $5\text{-}HT_4$ receptor-mediated stimulation of intestinal motility and secretion.[90]

Treatment of Functional Bowel Disorders

Functional bowel disorders are characterized by symptoms attributed to the mid or lower intestinal tract and for which no structural or biochemical cause can be ascribed.[91] The most common of this group of disorders is irritable bowel syndrome (IBS) in which abdominal pain is associated with a disturbed defecation pattern (alternating diarrhea/constipation).[91] Although the pathophysiology of this disorder is not clearly understood, a combination of disordered motor function and visceral sensitivity is believed to be present.[91]

A pathophysiological role of 5-HT and $5\text{-}HT_4$ receptors in IBS has been proposed based on the following grounds.[92,93] Two studies have reported elevated plasma levels of 5-HT in patients with abdominal pain and diarrhea, presumably related to IBS,[94,95] although a third study failed to show increased excretion of urinary metabolites of 5-HT.[96] Systemic administration of 5-hydroxytryptophan (5-HTP), a precursor of 5-HT, to human volunteers evokes IBS-like symptoms such as diarrhea and abdominal cramping.[97] In mice, 5-HTP-induced stimulation of watery diarrhea and defecation is antagonized by selective $5\text{-}HT_4$ receptor antagonists.[98,99] This may also be true in man since in vitro studies have demonstrated a $5\text{-}HT_4$ receptor-mediated stimulatory effect on smooth muscle tone[20] and enterocyte secretion in human intestine.[75-77] A role of $5\text{-}HT_4$ receptors in direct mediation of visceral pain transmission is unlikely since preclinical studies have shown no effect of selective $5\text{-}HT_4$ receptor antagonists in a rat model of colorectal distension-induced visceral pain.[100]

5-HT$_4$ receptor activation has been shown to sensitize sensory (AH-type) neurons by causing slow depolarization.[23] Additionally, 5-HT$_4$ receptor agonism at enteric interneurons (S-type) facilitates motor transmission.[23,24] In this manner, 5-HT$_4$ receptors are strategically located within the enteric nervous system, such that its overactivation can potentially lead to the heightened gut sensitivity (urge-to-defecate) that is typically noted in IBS patients. It is pertinent to note that, unlike selective 5-HT$_3$ receptor antagonists, selective 5-HT$_4$ receptor antagonists do not affect normal defecation.[99] Collectively, these observations suggest that selective 5-HT$_4$ receptor antagonists[101] may restore normal bowel function, particularly in diarrhea-predominant IBS patients. Also, it is conceivable that this class of drugs may alleviate abdominal pain, secondary to their ability to diminish intestinal hyperactivity.

Conclusions

About seven years have elapsed since the seminal discovery of functional 5-HT$_4$ receptors in the guinea pig ileum. During this period enormous strides have been made in our understanding of the role of 5-HT$_4$ receptors in regulating various aspects of gastrointestinal function. Importantly, several selective 5-HT$_4$ receptor agonists and antagonists have been developed which, besides serving as valuable pharmacological tools, may enable us to elucidate the physiological and pathophysiological significance of this receptor in the gastrointestinal tract. The next few years should witness several Phase II and III clinical trials with selective 5-HT$_4$ receptor agonists and antagonists, the outcomes of which are awaited with considerable interest.

References

1. Erspamer V. Occurrence of indolealkylamines in nature. In: Erspamer V, ed. Handbook of Experimental Pharmacology. Val 19: 5-Hydroxytrptamine and Related Indolealkylamines. New York: Springer-Verlag, 1966; 132-181.
2. Gershon MD, Wade PR, Kirchgessner AL, Tamir H. 5-HT receptor subtypes outside the central nervous system: Roles in the physiology of the gut. Neuropschopharmacol 1990; 3:385-395.
3. Dhasmana KM, Zhu YN, Cruz SL, Villalon CM. Gastrointestinal effects of 5-hydroxytryptamine and related drugs. Life Sc 1993; 53:1651-1661.
4. Furness JB, Costa M. Neurons with 5-hydroxytryptamine-like immunoreactivity in the enteric nervous system: Their projections in the guinea pig small intestine. Neuroscience 1982; 7:341-350.

5. Ford APDW, Clarke DE. The 5-HT$_4$ receptor. Med Res Rev 1993; 13:633-662

6. Hegde SS, Eglen RM. Peripheral 5-HT$_4$ receptors. FASEB J 1996; 10:1398-1407.

7. Craig DA, Clarke DE. Pharmacological characterization of a neuronal receptor for 5-hydroxytryptamine in guinea pig ileum with properties similar to the 5-hydroxytryptamine$_4$ receptor. J Pharm Exp Ther 1990; 252:1378-1386.

8. Uchiyama-Tsuyuki Y, Satoh M, Muramatsu M. Identification and characterization of the 5-HT$_4$ receptor in the intestinal tract and striatum of the guinea pig. Life Sc 1996; 59:2129-2137.

9. Gerald C, Adham N, Kao H-T, Olsen MA, Laz TM, Schechter LE, Bard JA, Vaysse PJJ, Hartig PR, Branchek TA, Weinshank RL. The 5-HT$_4$ receptor: molecular cloning and pharmacological characterization of two splice variants. EMBO J 1995; 14:2806-2815.

10. Ford APDW, Baxter GS, Eglen RM, Clarke DE. 5-Hydroxytryptamine stimulates cyclic AMP formation in the tunica muscularis mucosae of the rat oesophagus via 5-HT$_4$ receptors. Br J Pharmacol 1992; 211:117-120.

11. McLean PG, Coupar I.M. Stimulation of cyclic AMP formation in the circular smooth muscle of human colon by activation of 5-HT$_4$-like receptors. Br J Pharmacol 1996; 117:238-239.

12. Fagni L, Dumuis A, Sebben M, Bockaert J. The 5-HT$_4$ receptor subtype inhibits K$^+$ current in colluculi neurons via activation of a cyclic AMP-dependent protein kinase. Br J Pharmacol 1992; 105:973-979.

13. Craig DA, Eglen RM, Walsh LKM, Perkins LA, Whiting RL, Clarke DE. 5-Methoxytryptamine and 2-methyl-5-hydroxytryptamine-induced desensitisation as a discriminative tool for the 5-HT$_3$ and putative 5-HT$_4$ receptors in guinea pig ileum. Naunyn-Schmiedeberg Arch Pharmacol 1990; 342:9-16.

14. Ronde P, Ansanay H, Dumuis A, Miller R, Bockaert J. Homologous desensitization of 5-hydroxytrptamine$_4$ receptors in rat esophagus: Functional and second messenger studies. J Pharmacol Exp Ther 1995; 272:977-983.

15. McLean PG, Coupar IM, Molenaar P. Changes in sensitivity of 5-HT receptor mediated functional responses in the rat oesophagus, fundus and jejunum following chronic infusion with 5-hydroxytryptamine. Naunyn-Scmiedeberg Arch Pharmacol 1996; 354:513-519.

16. Buchheit K-H, Engel G, Mutschler E, Richardson B. Study of the contractile effect of 5-hydroxytryptamine (5-HT) in the isolated longitudinal muscle strip from guinea pig ileum. Evidence for two distinct release mechanisms. Naunyn-Scmiedeberg Arch Pharmacol 1985; 329:36-41.

17. Gaddum JH, Picarelli ZP. Two kinds of tryptamine receptor. Br J Pharmacol 1957; 12:323-328.

18. Tonini M, Candura SM, Onori L, Coccini T, Manzo L, Rizzi CA. 5-Hydroxytryptamine$_4$ receptors agonists facilitate cholinergic trans-

mission in the circular muscle of guinea pig ileum: antagonism by tropisetron and DAU 6285. Life Sci 1992; 50:PL173-178.

19. Eglen RM, Swank SR, Walsh LKM, Whiting RL. Characterization of 5-HT₃ and atypical 5-HT receptors mediating guinea pig ileal contractions in vitro. Br J Pharmacol 1990; 101:513-520.

20. Wardle KA, Sanger GJ. SB 207266 is a potent 5-HT₄ receptor antagonist in human isolated gastrointestinal tissue. Br J Pharmacol 1997; 120:192P.

21. Kilbinger H, Wolf D. Effects of 5-HT₄ receptor stimulation on basal and electrically evoked release of acetylcholine from guinea pig myenteric plexus. Naunyn-Schmiedeberg's Arch Pharmacol 1992; 345:270-275.

22. Kilbinger H, Gebauer A, Haas J, Ladinsky H, Rizzi CA. Benzimidazolones and renzapride faciliate acetylcholine release from guinea pig myenteric plexus via 5-HT₄ receptors. Naunyn-Schmiedeberg's Arch Pharmacol 1995; 351:229-236.

23. Tonini M, Galligan JJ, North RA. Effects of cisapride on cholinergic neurotransmission and propulsive motility in the guinea pig ileum. Gastroenterology 1989; 96:1257-1264.

24. Pan H, Galligan JJ. 5-HT₁A and 5-HT₄ receptors mediate inhibition and facilitation of fast synaptic transmission in enteric neurones. Am J Physiol 1994; 266:G230-238.

25. King BF, Sanger GJ. Facilitation of non-cholinerigc neurotransmission by renzapride (BRL 24924) in circular muscle of guinea pig ileum. Br J Pharmacol 1992; 107:313P.

26. Ramirez MJ, Cenarrazabeitia E, Del Rio J, Lasheras B Involvement of neurokinins in the non-cholinergic response to activation of 5-HT₃ and 5-HT₄ receptors in guinea pig ileum. Br J Pharmacol 1994; 111:419-424.

27. Elswood CJ, Bunce KT, Humphrey PPA. Identification of putative 5-HT₄ receptors in guinea pig ascending colon. Eur J Pharmacol 1991; 196:149-155.

28. Kojima S, Shimo Y. An enhancing effect of 5-hydroxytryptamine on electrically evoked, atropine resistant contraction of guinea pig proximal colon. Br J Pharmacol 1995; 114:73-76.

29. Kojima S, Shimo Y. Investigation into the 5-hydroxytryptamine-induced atropine-resistant neurogenic contraction of guinea pig proximal colon. Br J Pharmacol 1996; 117:1613-1618.

30. Briejer MR, Schuurkes JAJ. 5-HT₃ and 5-HT₄ receptors and cholinergic and tachykininergic neurotransmission in the guinea pig proximal colon. Eur J Pharmacol 1996; 308:173-180.

31. Wardle KA, Sanger GJ. The guinea pig distal colon—a sensitive preparation for the investigation of 5-HT₄ receptor mediated contractions. Br J Pharmacol 1993; 110:1593-1599.

32. Woolard DJ, Bornstein JC, Furness JB. Characterization of 5-HT receptors mediating contraction and relaxation of the longitudinal muscle of guinea pig distal colon in vitro. Naunyn-Schmiedeberg's Arch Pharmacol 1994; 349:455-462.

33. Hegde SS, Ku P, Lai K, Eglen RM. 5-HT$_4$ receptor mediated contraction of colon in the anaesthetized guinea pig. Br J Pharmacol 1994; 112:559P.

34. Prins NH, Briejer MR, Schuurkes JAJ. Characterization of the contraction to 5-HT in the canine colon longitudinal muscle. Br J Pharamacol 1997; 120:714-720.

35. Tam FS-F, Hillier K, Bunce KT. Characterization of the 5-hydroxytryptamine receptor type involved in inhibition of spontaneaous activity of human isolated colonic circular muscle. Br J Pharmacol 1994; 113:143-150.

36. Gebauer A, Merger M, Kilbinger H. Modulation by 5-HT$_3$ and 5-HT$_4$ receptors of the release of 5-hydroxytrptamine from the guinea pig small intestine. Naunyn-Schmiedberg's Arch Pharmacol 1993; 347:137-140

37. Minami M, Tamakai H, Ogawa T, Endo T, Hamaue N, Hirafuji M, Yoshioka M, Blower PR. Chemical modulation of 5-HT$_3$ and 5-HT$_4$ receptors affects the release of 5-hydroxytryptamine from the ferret and rat intestine. Res Comm Mol Path Pharmacol 1995; 89:131-142.

38. Briejer MR, Akkermans LMA, Shuurkes JAJ. Gastrointestinal prokinetic benzamides: The pharmacology underlying stimulation of motility. Pharmacol Rev 1995; 47:631-651.

39. Gullikson GA, Virina MA, Loeffler RF, Yang D-C, Goldstin B, Wang S-X, Moummi C, Flynn DL, Zabrowski DL SC 49518 enhances gastric emptying of solid and liquid meals and stimulates gastrointestinal motility in dogs by a 5-hydroxytryptmine$_4$ receptor mechanism. J Pharmacol Exp Ther 1993; 264:240-248.

40. Rizzi CA, Sagrada A, Schiavone A, Schiantarelli P, Cesana R, Schiavi GB, Ladinsky H, Donetti A. Gastroprokinetic properties of the benzimidazolone derivative BIMU 1, an agonist at 5-hydroxytryptamine$_4$ and an antagonist at 5-hydroxytryptmine$_3$ receptors. Naunyn-Schmiedeberg's Arch Pharmacol 1994; 349:338-345.

41. Bingham S, King BF, Rushant B, Smith MI, Gaster L, Sanger GJ. Antagonism by SB 204070 of 5-HT-evoked contractions in the dog stomach: an in vivo model of 5-HT$_4$ receptor function. J Pharm Pharmacol 1995; 47:219-222.

42. Schiavone A, Volonte M, Micheletti. The gastrointestinal motor effect of benzamide derivatives is unrelated to 5-HT$_3$ receptor blockade. Eur J Pharmacol 1990; 187:323-329.

43. Cohen ML, Bloomquist W, Gidda JS, Lacefield W. LY277359 maleate: A potent and selective 5-HT$_3$ receptor antagonist without gastroprokinetic activity. J Pharmacol Exp Ther 1990; 254:350-355.

44. Hegde SS, Wong AG, Perry MR, Ku P, Moy TM, Loeb M, Eglen RM. 5-HT$_4$ receptor mediated stimulation of gastric emptying in rats. Naunyn-Schmiededberg's Arch Pharmacol 1995; 351:589-595.

45. Gale JD, Reeves JJ, Bunce KT. Antagonism of the gastroprokinetic effect of metoclopramide by GR125487, a potent and selective 5-HT$_4$ receptor antagonist. J Gastrointest Motil 1994; 5:192.

46. Amemiya N, Hatta S, Takemura H, Ohshika H. Characterization of the contractile response induced by 5-methoxytryptamine in rat stomach fundus strips. Eur J Pharmacol 1996; 318:403-409.

47. Buchheit K-H, Buhl T. Stimulant effects of 5-hydroxytryptamine on guinea pig stomach preparation in vitro. Eur J Pharmacol 1994; 262: 91-97.

48. Matsuyama S, Sakiyama H, Nei K, Tanaka C. Identification of putative 5-hydroxytrptamine$_4$ (5-HT$_4$) receptors in guinea pig stomach: The effect of TSK159, a novel agonist, on gastric motility and acetylcholine release. J Pharmacol Exp Ther 1996; 276:989-995.

49. Schuurkes JAJ, Meulmans AL, Obertop H, Akkermans LMA. 5-HT$_4$ receptors on human stomach. J Gastrointes Motil 1991; 3:P199.

50. Plaza MA, Arrueba MP, Murillo MD. 5-hydroxytryptamine induces forestomach hypomotility in sheep through 5-HT$_4$ receptors. Exp Physiol 1996; 81:781-790.

51. Reeves JJ, Bunce KT, Humphrey, PPA. Investigation into the 5-hydroxytryptamine receptor mediating smooth muscle relaxation in the rat oesophagus. Br J Pharmacol 1991; 103:1067-1072.

52. Baxter GS, Craig DA, Clarke DE. 5-Hydroxytryptamine$_4$ receptors mediated relaxation of the rat oesophageal tunica muscularis mucosae. Naunyn-Schmiedeberg's Arch Pharmacol 1991; 343:439-446.

53. Tuladhar BR, Costall B, Naylor RJ. Pharmacological characterization of the 5-hydroxytryptamine receptor mediating relaxation in the rat isolated ileum. Br J Pharmacol 1996; 119:303-310.

54. Waikar MV, Hegde SS, Ford APDW, Clarke DE. Pharmacological analyzes of Endo-6-methoxy-8-methyl-8-azabicyclo[3.2.1]oct-3-yl-2,3-dihydo-2-oxo-1H-benzimidazole-1-carboxylate hydrochloride (DAU 6285) at the 5-hydroxytryptamine$_4$ receptor in the tunica muscularis mucosae of rat esophagus and ileum of guinea pig: Role of endogenous 5-hydroxytryptamine. J Pharmacol Exp Ther 1993; 264:654-661.

55. Tuladhar BR, Costall B, Naylor RJ. Ability of isobutylmethyl xanthine to potentiate the 5-HT$_4$ receptor mediated relaxant effect of both endogenous and exogenous 5-HT in the rat ileum. Br J Pharmacol 1994; 113:118P.

56. Tam FS-F, Hillier K, Bunce KT, Grossman C. Differences in response to 5-HT$_4$ receptor agonists and antagonists of the 5-HT$_4$-like receptor in human colon circular smooth muscle. Br J Pharmacol 1995; 115:172-176.

57. McLean PG, Coupar IM, Molenaar P. A comparative study of functional 5-HT$_4$ receptors in human colon, rat oesophagus and rat ileum. Br J Pharmacol 1995; 115:47-56.

58. Costa M, Furness JB. The peristaltic reflex: An analysis of the nerve pathways and their pharmacology. Naunyn-Schmiedeberg's Arch Pharmacol 1976; 294:47-60.

59. Craig DA, Clarke DE. Peristalsis evoked by 5-HT and renzapride: evidence for putative 5-HT$_4$ receptor activation. Br J Pharmacol 1991; 102:563-564.

60. Buchheit K-H, Buhl T. Prokinetic benzamides stimulate peristaltic activity in the isolated guinea pig ileum by activation of 5-HT$_4$ receptors. Eur J Pharmacol 1991; 205:203-208.

61. Costall B, Naylor RJ, Tuladhar BR. 5-HT$_4$ receptor mediated facilitation of the emptying phase of the peristaltic reflex in the guinea pig isolated ileum. Br J Pharmacol 1993; 110:1572-1578.

62. Tuladhar BR, Costall B, Naylor RJ. 5-HT$_3$ and 5-HT$_4$ receptor mediated facilitation of the emptying phase of the peristaltic reflex in the marmoset isolated ileum. Br J Pharmacol 1996; 117:1679-1684.

63. Yuan SY, Bornstein JC, Furness JB. Investigation of the role of 5-HT$_3$ and 5-HT$_4$ receptors in ascending and descending reflexes to the circular muscle of the guinea pig small intestine. Br J Pharmacol 1994; 112:1095-1100.

64. Kadowaki M, Wade PR, Gershon MD. Participation of 5-HT$_3$, 5-HT$_4$, and nicotinic receptors in the peristaltic reflex of guinea pig distal colon. Am J Physiol 1996; 34:G849-G857.

65. Grider JR, Kuemmerle JF, Jin J-G. 5-HT released by mucosal stimuli initiates peristalsis by activating 5-HT$_4$/5-HT$_{1P}$ receptors on sensory CGRP neurons. Am J Physiol 1996; 33:G778-G782.

66. Foxx-Orenstein AE, Kuemmerle JF, Grider JR. Distinct 5-HT receptors mediate the peristaltic reflex induced by mucosal stimuli in human and guinea pig intestine. Gastroenterol 1996; 111:1281-1290.

67. Grider JR, Jin J-G. Distinct populations of sensory neurons mediate the peristaltic reflex elicited by muscle stretch and mucosal stimulation. J Neurosc 1994; 14:2854-2860.

68. Leung EL, Blissard D, Jett MF, Eglen RM. Investigation of the 5-hydroxytryptamine receptor mediating the "transient" short-circuit current response in guinea pig ileal mucosa. Naunyn-Schmiedeberg's Arch Pharmacol 1995; 351:596-602.

69. Scott CM, Bunce KT, Spraggs CF. Investigation of the 5-hydroxytryptamine receptor mediating the "maintained" short-circuit current response in guinea pig ileal mucosa. Br J Pharmacol 1992; 106:877-882.

70. Johnson PJ, Bornstein JC, Furness JB, Woollard DJ, Ormann-Rossiter SL. Characterization of 5-hydroxytryptamine receptors mediating mucosal secretion in guinea pig ileum. Br J Pharmacol 1994; 111:1240-1244.

71. Grossman CJ, Bunce KT. 5-HT$_3$ and 5-HT$_4$ receptors mediate the secretory effects of 5-HT in rat distal colon. Br J Pharmacol 1994; 112:176P.

72. Franks CM, Hardcastle J, Hardcastle PT, Sanger GJ. Do 5-HT$_4$ receptors mediate the intestinal secretory response to 5-HT in rat in vivo. J Pharm Pharmacol 1994; 47:213-218.

73. Hansen MB, Jaffe BM. 5-HT receptor subtypes involved in luminal serotonin-induced secretion in rat intestine in vivo. J Surg Res 1994; 56:277-287.

74. Hansen MB. ICS 205-930 reduces 5-methoxytryptamine-induced short circuit current in stripped pig jejunum. Can J Physiol Pharmacol 1994; 72:227-232.

75. Burleigh DE, Borman RA. Short-circuit current responses to 5-hydroxytryptamine in human ileal mucosa are mediated by a 5-HT$_4$ receptor. Eur J Pharmacol 1993; 241:125-128.

76. Budhoo HR, Harris RP, Kellum JM. The role of the 5-HT$_4$ receptor in Cl⁻ secretion in human jejunal mucosa. Eur J Pharmacol 1996; 314:109-114.

77. Borman RA, Burleigh DE. Human colonic mucosa possesses a mixed population of 5-HT receptors. Eur J Pharmacol 1996; 309:271-274.

78. Rhodes KF, Lattimer N, Coleman J. A component of 5-HT evoked depolarization of rat isolated vagus nerve is mediated by a putative 5-HT$_4$ receptor. Naunyn-Schmiedeberg's Arch Pharmacol 1992; 346:496-503.

79. Bley K, Eglen RM, Wong, EHF. Characterization of 5-hydroxytryptamine-induced depolarizations in rat isolated vagus nerve. Eur J Pharmacol 1994; 260:139-147.

80. Fukui H, Yamamoto M, Sasaki S, Sato S. Possible involvement of peripheral 5-HT$_4$ receptors in copper sulfate induced vomiting in dogs. Eur J Pharmacol 1994; 257:47-52.

81. Bhandari P, Andrews PLR. Preliminary evidence for the involvement of the putative 5-HT$_4$ receptor in zacopride and copper sulphate-induced vomiting in the ferret. Eur J Pharmacol 1991; 204:273-280.

82. Qin XY, Pilot MA, Thompson H, Scott M. Effects of cholinoceptor and 5-hydroxytrypyamine$_3$ receptor antagonism on erythromycin-induced canine intestinal motility disruption and emesis. Br J Pharmacol 1993; 108:44-49.

83. Tonini M. Recent advances in the pharmacology of gastrointestinal prokinetics. Pharmacol Res 1996; 33:217-226.

84. Turconi M, Schiantarelli P, Borsini F, Rizzi CA, Ladinsky H, Donetti A. Azabicycloalkyl benzimidazolones: interaction with serotonergic 5-HT$_3$ and 5-HT$_4$ receptors and potential therapeutic implications. Drugs Future 1991; 16:1011-1026.

85. Garcia-Garayoa E, Monge A, Roca J, Del Rio J, Lasheras B. VB20B7, a novel 5-HT-ergic agent with gastrokinetic activity. II. Evaluation of the gastroprokinetic activity in rats and dogs. J Pharm Pharmacol 1997; 49:66-73.

86. Wiseman LR, Faulds D. Cisapride. An updated review of its pharmacology and therapeutic efficacy as a prokinetic agent in gastrointestinal motility disorders. Drugs 1994; 47:116-152.

87. Croci T, Langlois M, Mennini T, Landi M, Manara L. ML 10302, a powerful and selective new 5-HT$_4$ receptor agonist. Br J Pharmacol 1995; 114:382P.

88. Clark RD, Jahangir A, Langston JA, Weinhardt KK, Miller B, Leung E, Eglen RM. Ketones related to the benzoate 5-HT$_4$ receptor antagonist RS 23597 are high affinity partial agonists. Bioorg Med Chem Lett 1994; 4:2477-2480.

89. Buchheit K-H, Gamse R, Giger R, Hoyer D, Klein F, Kloppner E, Pfannkuche H-J, Mattes H. The serotonin 5-HT$_4$ receptor. 2. Struc-

ture-activity studies of the indole carbazimidamide class of agonists. J Med Chem 1995; 38:2331-2338.

90. Appel S, Kumle A, Hubert M, Duvauchelle T. First pharmacokinetic-pharmacodynamic study in humans with a selective 5-hydroxytryptamine$_4$ receptor agonist. J Clin Pharmacol 1997; 37:229-237.

91. Thompson WG, Gick M. Irritable bowel syndrome. Sem Gastrointes Dis 1996; 7:217-229.

92. Talley NJ. Review article: 5-hydroxytryptamine agonists and antagonists in the modulation of gastrointestinal motility and sensation: clinical implications. Alim Pharmacol Ther 1992; 6:273-289.

93. Sanger GJ. 5-hydroxytryptamine and functional bowel disorders. Neurogastroenterol Mot 1996; 8:319-331.

94. Warner RRP. Hyperserotoninemia in functional gastrointestinal disease. Ann Int Med 1963; 59:464-476.

95. Shimizudani T. Changes in the blood serotonin levels in cases of upper gastrointestinal tract diseases. Tokyo Ika Diagaku 1973; 31:713-733.

96. Kowlessar OD, Williams RC, Law DH, Sleisenger MH. Urinary excretion of 5-hydroxyindoleacetic acid in diarrhea states, with special reference to nontrupical sprue. N Eng J Med 1958; 259:340-341.

97. Davidson JD, Sjoerdsma A, Loomis LN, Udenfriend S. The effects of 5-hydroxytryptophan, the precursor of serotonin, in experimental animals and man. Clin Res Proc 1957; 5:304.

98. Hegde SS, Moy TM, Perry MR, Loeb M, Eglen RM. Evidence for the involvement of 5-hydroxytryptamine$_4$ receptors in 5-hydroxytryptophan-induced diarrhea in mice. J Pharmacol Exp Ther 1994; 271:741-747.

99. Banner SE, Smith MI, Bywater D, Gaster LM, Sanger GJ. Increased defaecation caused by 5-HT$_4$ receptor activation in the mouse. Eur J Pharmacol 1996; 304:181-186.

100. Hegde SS, Bonhaus DW, Johnson LG, Leung E, Clark RD, Eglen RM. RS 39604: a potent, selective and orally active 5-HT$_4$ receptor antagonist. Br J Pharmacol 1995; 115:1087-1095.

101. Gaster LM, King FD. Serotonin 5-HT$_3$ and 5-HT$_4$ receptor antagonists. Med Res Rev 1997; 17:163-214.

CHAPTER 8

5-HT$_4$ Receptors in Lower Urinary Tract Tissues

Anthony P.D.W. Ford and M. Shannon Kava

Introduction

Interest in the role of 5-HT in the physiology and pathophysiology of the lower urinary tract (LUT), and its effect on neurotransmission in the bladder and urethra, has developed considerably over the last 20 years,[1,2] reflecting the increased awareness of therapeutic potential and the relative inadequacy of current therapies for LUT disorders. 5-HT is recognized now as an ubiquitous intercellular signaling molecule. A great diversity of cell types can synthesize, store, release, uptake, as well as respond to extracellular 5-HT. This, coupled with the tremendous number and diversity of receptor proteins, implicates 5-HT in many physiological and pathophysiological processes. The importance of 5-HT in the regulation of the behavior of the "hollow organs" (e.g., discrete parts of the alimentary tract, the heart, the vasculature, and the bladder) has been well established.[3-5] It is now clear that the many receptors for 5-HT (at least 14 known subtypes) allow for so many forms of modulation that tremendous subtleties in organ function can be introduced by selective receptor subtype activation or antagonism. Largely because 5-HT functions as a modulatory, rather than direct, transmitter in so many organs, the attractiveness of selective interaction with 5-HT receptors is great. Clearly, the opportunities for new, selective disease therapies are tremendous, and diseases of the LUT, which are poorly served by current drugs, are a key area for advancement in an ever-aging population.

5-HT$_4$ Receptors in the Brain and Periphery, edited by Richard M. Eglen.
© 1998 Springer-Verlag and R.G. Landes Company.

In this chapter, the role played by 5-HT in the LUT, specifically via action at the 5-HT$_4$ receptor subtype, is detailed and reviewed. By necessity, and for reasons that will soon become apparent, the focus is placed on the human urinary bladder. Attention is given to several particularly important questions: At what sites and by what mechanisms can 5-HT act? How important is species dependence? What may happen in disease processes? How may LUT function be influenced (or improved) by 5-HT$_4$ receptor modulation? What are the LUT sources of 5-HT? It is assumed that 5-HT$_4$ receptors are, pharmacologically and functionally, a homogeneous population, despite the description of splicing variants (5-HT$_{4L}$ and 5-HT$_{4S}$).[6] After a discussion of these points, studies performed in our laboratory on bladder tissues from monkey and man are described and compared, to provide examples of the species-dependent nature of 5-HT$_4$ receptor-mediated events in the bladder, and to emphasize the significance of the modulation seen in human bladder.

5-HT Responses in LUT Tissues: Species Dependence

Pharmacological studies on the urinary bladder have demonstrated great species variation with regard to the nature of responses to 5-HT and the subtype(s) of 5-HT receptor involved.[2] For example, in dog, 5-HT contracts the isolated bladder via 5-HT$_2$ receptor subtypes.[7] 5-HT-induced contractions of the isolated bladder of guinea-pig[8] are reportedly insensitive to methysergide (a "nonselective" 5-HT$_1$ and 5-HT$_2$ receptor antagonist). In urinary bladder of anesthetized cat, biphasic excitatory dose-response curves to 5-HT have been reported and are mediated via 5-HT$_3$ and 5-HT$_2$ receptors.[9] In general, direct (i.e., non-neurogenic) contractile responses to 5-HT have been demonstrated in whole bladder or strips of detrusor from most species studied, including rat, rabbit, guinea pig, dog, pig, monkey and man.[2] In many species, a 5-HT$_2$ receptor subtype has been implicated. However, in a clinical study with ketanserin, a selective 5-HT$_{2A}$ receptor antagonist, no improvement in bladder compliance was observed in patients with detrusor instability, ruling out a role for this receptor subtype in the pathophysiology of urge incontinence.[1] It is relatively clear, and also highly likely (given that 5-HT$_4$ receptors couple preferentially via the G$_s$ G-protein and stimulation of intracellular adenylyl cyclase), that no direct, excitatory smooth muscle action of 5-HT is mediated by 5-HT$_4$ receptor activation.

In the context of excitatory responses to electrical stimulation (i.e., neurogenic stimulation), greater potential exists for modulatory serotonergic mechanisms, and more 5-HT receptor subtypes may be involved. In one study, using superfused mouse bladder strips, responses to electrical stimulation were potentiated by 5-carboxamidotryptamine (5-CT) with greater potency than 5-HT and N-omega-methyl-5-HT, in turn more potent than α-methyl-5-HT.[10] The effect of 5-HT was resistant to blockade with conventional 5-HT antagonists, including ketanserin, methysergide, methiothepin, and MDL 72222, even at concentrations as high as 1 μM. Thus, most 5-HT receptors defined at that time were essentially eliminated from involvement (5-HT_1, 5-HT_2 and 5-HT_3 subtypes). Evidence to suggest involvement of the 5-HT_{1B} receptor was described, on the basis of sensitivity to pindolol. Confirmation of this proposal was forwarded in a separate study, which demonstrated that potentiation of electrically-evoked contractions by 5-HT could be mediated through 5-HT_{1B} and 5-HT_2 receptors.[11] Whether this characterization stands in the light of newly-defined receptors (e.g., 5-HT_6 and 5-HT_7) which couple through pathways hypothetically more consistent with presynaptic facilitation (adenylyl cyclase activation, unlike the 5-HT_{1B} receptor), remains to be seen.

In terms of inhibitory responses to 5-HT, fewer examples have been reported from studies of the effect of 5-HT on responses to electrical stimulation of intramural nerves in the bladder, or on responses to direct stimulation of the detrusor. Again, there is a diversity of receptor involvement and considerable species dependence is illustrated. 5-HT relaxes isolated bladder neck of pig, an effect blocked by methysergide (10 μM; an antagonist at 5-HT_1 and 5-HT_2 receptors), but not ketanserin (1 μM; a selective 5-HT_{2A} receptor antagonist).[12] In bladder of bullfrog, 5-HT-induced inhibition of electrically-evoked contractions is insensitive to methysergide.[13] In a single study[14] an inhibitory response to 5-HT mediated at the postjunctional level was attributed to activation of the 5-HT_4 receptor. It remains to be seen if such a mechanism may also apply in other species, particularly man.

In Vitro Evidence for 5-HT_4 Receptors in LUT of Man

Pharmacological characterization of 5-HT receptors in urinary bladder of man has, until the last few years, been hampered not only by the availability of tissue, but by a lack of specific ligands.[15] The most convincing, early evidence for a putative 5-HT$_4$ receptor can be found in the work of Hindmarsh et al[16] who reported potentiation of electrically-induced contractions of human isolated bladder by "remarkably low" concentrations of 5-HT (as low as 0.1 nM), an effect insensitive to blockade by methysergide and morphine. With the benefit of our own experimental data, generated over the last 3-4 years, these potentiations of electrically-induced contractions by low concentrations of 5-HT described by Hindmarsh are almost entirely consistent with activation of prejunctional 5-HT$_4$ receptors. It is noteworthy that their study included the observation that the potentiation by 5-HT was observed even in the presence of atropine. This implies that both acetylcholine-mediated and -independent (NANC) transmission from the parasympathetic efferent fibers, which we know now to be carried largely by ATP, are modulated by this mechanism.

The early data from Hindmarsh and co-workers remained largely hidden for over a decade, until in 1991, Corsi et al described an "atypical" 5-HT receptor in human bladder that mediated potentiation of contractile responses to electrical field stimulation.[17] The facilitation with 5-HT was not seen with responses to direct muscle stimulation (using stimuli of broader pulse width, in the presence of tetrodotoxin) or following administration of a spasmogen (acetylcholine), indicating that 5-HT probably interacted with prejunctional parasympathetic terminals or intramural ganglia. Elements of the agonist and antagonist profile for this response resemble that of the 5-HT$_4$ receptor, including potent agonism by 5-HT ($pEC_{50} = 8.0$) and 5-methoxytryptamine (5-MeOT), agonism by certain substituted benzamide derivatives, antagonism by micromolar concentrations of tropisetron (ICS 205-930), and resistance to inhibition by certain 5-HT$_1$, 5-HT$_2$ and 5-HT$_3$ receptor antagonists.[4,18] However, antagonism by tropisetron in human bladder was complex and at high concentrations tropisetron behaved as an agonist.[17] Furthermore, the potentiating action of 5-HT was antagonized, in part, by methysergide (1 µM). There was clearly adequate evidence that a 5-HT$_4$ receptor mechanism operated in their initial studies; however, either the characteristics of the human bladder 5-HT$_4$ receptor were different, or failure to isolate pharmacologically a singular receptor component confused the characteristics.

Tonini and coworkers[19] extended the observations of Corsi,[17] taking advantage of the availability of a high-affinity, selective 5-HT_4 receptor antagonist, GR 113808.[20] In these studies of field-stimulated detrusor strips of man (from the anterior part of the bladder dome), GR 113808 (3-30 nM) produced concentration-dependent, parallel displacements of 5-HT concentration-effect curves, and an affinity estimate (pA_2) of 8.9 confirmed the involvement of a 5-HT_4 receptor. A key methodological change in their study was the addition of methysergide and ondansetron to the Krebs buffer (see ref. 14), in order to isolate pharmacologically the 5-HT_4 receptor from any 5-HT_1, 5-HT_2 or 5-HT_3 receptors that may complicate the responses. On this evidence, the authors proposed that selective 5-HT_4 receptor agonists could be valuable for treatment of detrusor hypomotility conditions. Authors from the same laboratory followed up on these findings[21] with a thorough characterization of the 5-HT_4 receptor population present in isolated detrusor of man, cataloguing the effects of a range of indole, substituted benzamide and benzimidazoline agonists and antagonists. These studies, using strips of anterior dome of bladder from cystectomy patients, were again performed in the presence of methysergide and ondansetron. The agonist profile that emerged from these studies was essentially typical for 5-HT_4 receptors, although the potency of 5-methoxytryptamine (100 times lower than 5-HT) was rather weaker than might be expected, possibly reflecting a species difference of the human 5-HT_4 receptor or the possible amine oxidation of low concentrations of this agonist during the studies. Cisapride was shown to be a partial agonist in these studies, with approximately 50% of the intrinsic activity of 5-HT. Other antagonists tested to complete the 5-HT_4 receptor profile were GR 125487[22] ($pA_2 = 9.7$), DAU 6285[23] ($pA_2 = 6.4$), and the weak partial agonist, RS-23597[24] (approximate $pA_2 = 8.0$). The proposed mechanism by which 5-HT_4 receptor activation augments detrusor activity is depicted in Figure 8.1.

In Vitro Evidence for 5-HT_4 Receptors in Nonhuman LUT

In two other studies, investigators have proposed that a 5-HT_4 receptor population functions in the urinary bladder of a nonhuman mammal. In the first of these, Waikar et al[14] investigated whether the then "putative" 5-HT_4 receptor-mediating facilitation of parasympathetic stimulation in human detrusor, described in the studies of Corsi,[17] displayed similar location and function in two species

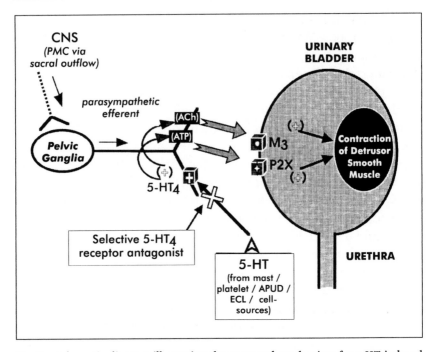

Fig. 8.1. Schematic diagram illustrating the proposed mechanism for 5-HT-induced enhancement of detrusor activity in bladder of man. 5-HT or selective agonists (e.g., cisapride, metoclopramide) activate $5\text{-}HT_4$ receptors located on postganglionic parasympathetic nerve terminals or on cells bodies (not shown) of intramural parasympathetic ganglia. $5\text{-}HT_4$ receptor activation leads to enhanced release of transmitters, principally ACh and ATP, which act on specific muscarinic (M_3 and M_2) and nucleotide (P2X1) receptors, respectively, on smooth muscle membranes to initiate contraction.

of monkey from the Macaque family. In these studies, which are described in more detail below, strips of the posterior aspect of the bladder dome from *Cynomolgus* and *Rhesus* monkeys were studied, in the presence of methysergide and ondansetron to favor the isolation of a possible $5\text{-}HT_4$ receptor component. Surprisingly, these investigators found that while 5-HT was able to modulate tetrodotoxin- and lignocaine-sensitive responses to electrical field-stimulation, the effect was an attenuation, rather than augmentation of responses. Furthermore, when studying the effects of direct electrical (broader pulse width) or chemical (carbachol) stimulation of the smooth muscle cells (in the presence of tetrodotoxin and lignocaine), 5-HT continued to exert an inhibitory effect, which bore the pharmacology of the $5\text{-}HT_4$ receptor (agonism by 5-methoxytryptamine and renzapride and sensitivity to antagonism by DAU 6285, GR 113808 and RS-23597-190). Consequently, the 5-HT$_4$ receptor population in the bladder of these Macaque monkeys was postulated as being located on the detrusor smooth muscle cells, rather than the parasympathetic nerves.

In guinea pig bladder strips, Messori et al[25] investigated 5-HT modulation of field stimulation responses and observed an apparently biphasic facilitation of responses. The low potency phase, responsible for approximately 70% of the total facilitation, was attributed to 5-HT$_3$ receptor stimulation, as it was sensitive to granisetron (0.3 μM). However, the potent phase (10-300 nM 5-HT) was slightly sensitive to inhibition by methiothepin, ketanserin and GR 125487, suggesting the involvement of two or more receptors, the 5-HT$_4$ receptor and, possibly, the 5-HT$_{2A}$. Of particular interest was the observation that hyoscine-resistant, suramin-sensitive components of the response to electrical stimulation were sensitive to the lower concentrations of 5-HT, and vice versa. Unfortunately, these investigations were not extended to characterize pharmacologically the potent phase after removal of the cholinergic component with hyoscine, which may have permitted better examination of the receptors involved. At best, these suggestions of 5-HT$_4$ receptor involvement in guinea pig bladder are intriguing, but tentative.

In Vivo Evidence for 5-HT$_4$ Receptors in LUT of Man

In reviewing the evidence for a role of the 5-HT$_4$ receptor in the regulation of the activity bladder of man, it is possible to pursue several different lines of research in the literature, each bearing varying degrees of relevance and significance. The most direct and psychologically satisfying approach is to study the activity of highly potent and selective 5-HT$_4$ receptor antagonists, with particular attention to bladder storage and voiding behavior, in normal volunteers, and in patients with a variety of presenting symptoms of urinary disorder (e.g., detrusor instability, interstitial cell cystitis, stress incontinence and chronic outlet obstruction). The ability to combine this approach with both subjective (symptomatic improvement, quality of life) and objective (filling cystometry, involuntary leakage occurrence) outcomes would be especially beneficial. Clearly, this would represent the completion of Phase I and II clinical trials for a new drug candidate in development for such an indication. As no data has been published to date from any clinical studies (volunteers or patients) using selective 5-HT$_4$ receptor antagonists, this ultimately satisfying information will be awaited with considerable interest.

In the absence of such information, the next best approach is to scour the literature for reports on the LUT-related activities of 5-HT_4 receptor agonists, which will provide information on the functional properties of the 5-HT_4 receptor, and the consequences of its activation, but will not attest to any endogenous, tonic role imparted by 5-HT itself. This approach is more fruitful, currently, since 5-HT_4 receptor agonists have in fact been used clinically for several years. The first of these, the substituted benzamide metoclopramide (Reglan), has been in clinical use for many years, and since well before its principal mechanism of action, 5-HT_4 receptor agonism, was elucidated. It is not without other pharmacological actions, possessing high 5-HT_3 and D_2 receptor affinity in addition to low- to- moderate efficacy at the 5-HT_4 receptor. Cisapride (Propulsid), also a substituted benzamide, has recently gained considerable clinical usage. It too displays other pharmacological actions, possessing moderate 5-HT_2 and D_2 receptor affinity combined with moderate to high 5-HT_4 receptor efficacy. Both metoclopramide and cisapride are indicated principally for management of gastrointestinal hypomotility disorders, such as reflux esophagitis, gastroparesis, and, especially after neurological impairment, for chronic constipation.[26,27]

The use of these agents has increasingly been associated with reports of urinary bladder effects that reflect a possible influence on the parasympathetic control of detrusor activity. The first such reference, in a patient case report, described the apparent reversal of diabetic neurogenic bladder.[28] Treatment with metoclopramide (10 mg/kg, i.v., 30 min before meals), in addition to improving motility in the alimentary canal, permitted spontaneous voiding in the individual, obviating the need for intermittent catheterization and reducing postvoid residuals to ~200 ml. After discontinuation of metoclopramide, spontaneous micturition ceased and catheterization yielded 960 ml urine. Restarting the metoclopramide allowed for complete reappearance of voiding. This observation was shortly followed by a letter briefly describing the case of a 55-year-old man receiving 10 mg metoclopramide orally for the treatment of gastroesophageal reflux, following which he described symptoms of urinary incontinence.[29] The authors surmised that the treatment had weakened urethral sphincter tone, although an effect directly on the parasympathetic outflow in the detrusor is equally plausible.

Shortly after the introduction of cisapride in Europe, a Scottish study investigated the ability of cisapride (10 mg, t.d.) to reduce colonic transit time in chronically constipated patients with paraplegia from spinal injury.[30] In addition to the positive findings regarding intestinal function, in seven out of ten patients, a reduction of residual urine volume was observed, while in one patient, discontinuation of cisapride led to acute urinary retention.

More recently, a report appeared from the Adverse Drug Reactions Advisory Committee of Australia,[31] in which 12 out of 86 reports of suspected adverse reactions associated with cisapride use described disorders of LUT function, including 5 involving urinary incontinence. In 8, urinary frequency was reported, as well as cystitis, hesitancy and acute retention. In all cases, urinary problems subsided after discontinuation of therapy. The authors speculated that cisapride, probably by stimulating cholinergic neurones in the bladder, causes detrusor contraction or instability. A separate report appeared the same year, again reporting increases in micturition frequency in 25 patients treated with cisapride.[32] In all but one of the cases where outcome was known, the urinary problems were relieved after withdrawal from cisapride.

Focusing on the potential for exploiting positively this apparent adverse reaction, Franceschetti et al[33] described how a 32-year-old woman with lazy bladder was treated with cisapride (10 mg, q.d.) for three weeks. A progressive improvement of symptoms and urodynamic parameters was seen, with a return to pretreatment conditions after drug withdrawal. This further supports the proposal that cisapride potentiates the release of acetylcholine from postganglionic parasympathetic nerves in the human detrusor by activating 5-HT$_4$ receptors, i.e., through the same mechanism responsible for its gastrointestinal prokinetic action. It was concluded that selective 5-HT$_4$ agonists could be potentially useful to improve bladder emptying in micturition disorders associated with detrusor hypocontractility.

This potential for the use of selective 5-HT$_4$ receptor agonists in disorders associated with partial or total detrusor failure was addressed by the same authors in a brief letter.[34] The currently accepted therapy for such disorders rests principally in clean, intermittent, self-catheterization. Drugs used previously to increase detrusor function, such as muscarinic partial agonists (e.g., bethanechol) or cholinesterase inhibitors (e.g., neostigmine) are poorly tolerated due to widespread side-effects and thus are seldom used. However, the effectiveness of self-catheterization is somewhat tempered by the unpleasantness and discomfort, potential for urethral and sphincter damage, and it is frequently a source for infectious contamination. In this light, the attractiveness of using selective 5-HT_4 receptor agonists, such as cisapride, is clear, with the potential for unpleasant side-effects being quite low.

A Role for Endogenous 5-HT in LUT Function?

The foregoing discussion establishes quite clearly the functional presence and significance of the 5-HT_4 receptor in the urinary bladder of man, as well as some therapeutic importance. What is lacking from such data, however, is any allusion to the role that may be played by endogenous agonist (5-HT itself) in modulating bladder activity in normal and diseased subjects. In the absence of data using selective 5-HT_4 receptor antagonists, one avenue worth pursuing is to review the literature from clinical use of agents which potentiate the action of endogenous 5-HT. Data from clinical experience with the widely prescribed serotonin-selective reuptake inhibitors (SSRIs) is particularly appropriate, not least because of the millions of prescriptions issued for agents such as fluoxetine, sertraline, paroxetine and venlafaxine (inter alia), in addition to the older, somewhat serotonin-selective tricyclic antidepressants.

One particularly noteworthy paper[35] reported a study of patients with neurogenic bladder dysfunction, some with and some without voluntary detrusor control or bladder sensation. The investigators chose to study the effects of 5-HT on bladder activity, not by introducing 5-HT itself, but by administering what was then the most serotonin-selective drug in use, clomipramine (75 mg, daily for 4 days). In this study, the intravesical volume, at which first detrusor reflex activity was observed, showed a decrease from 329 to 178 ml after clomipramine therapy. Also, residual urine volume decreased from 111 to 30 ml. Such information has to be considered in light of the accepted, if rare, use of other tricyclic agents (e.g., amitryptilline) in treating certain disorders associated with bladder hyperactivity and nocturia, as well as the known antimuscarinic properties of the tricyclic antidepressants, which would thus be expected yield greater, rather than reduced, intravesical and postvoid residual volumes. As far as the newer SSRIs are concerned, urinary incontinence and micturition frequency are listed among their adverse reaction profiles, although incidence from clinical studies appears to be rare.[36] One example from routine clinical practice has appeared: Sancho et al[37] recently described symptoms of pollakiuria (frequent voiding) and "micturition syndrome" (essentially, urgency) in patients treated with fluoxetine (Prozac). Whether any of these observations reflects changes exerted directly by altered 5-HT levels within the bladder, let alone via activation of 5-HT$_4$ receptor, is a matter for speculation.

What Potential Sources of 5-HT Exist in LUT Tissues?

A critical question that remains to be answered relates to the significance of tonic physiological and pathophysiological contributions of 5-HT to the function of the bladder and other LUT tissues. The clearest answer to this question will have to await clinical studies in normal volunteers and patients with urinary disorders using selective 5-HT$_4$ receptor antagonists. Nevertheless, the potential for use of selective 5-HT$_4$ receptor antagonists in urinary disorders is based on the hypothetical assertion that 5-HT is contributing to either the normal or disordered behavior of the bladder, via an action on parasympathetic ganglia or varicosities. Clearly, it is difficult to estimate how much 5-HT is required endogenously to influence parasympathetic outflow to the detrusor. It is evident from several of the studies described above that the 5-HT$_4$ receptor population present in human bladder is highly sensitive to 5-HT, indicating that a high receptor reserve exists in this tissue. In the studies of Hindmarsh et al,[16] Corsi et al[17] and in our own studies (see below), nanomolar, and occasionally subnanomolar concentrations of 5-HT can facilitate transmission. Indeed, in strips of urinary bladder from one patient, a young spina bifida patient undergoing augmentation surgery, we observed that concentrations as low as 3 pM produced facilitation, and the potency (pEC$_{50}$) of 5-HT was approximately 10. Thus, concentrations of 5-HT below the normal level of detection may be sufficient.

A highly speculative possibility is that the high efficiency of coupling of 5-HT$_4$ receptors on bladder parasympathetic efferents may be associated with a degree of 5-HT$_4$ receptor-G$_s$-protein precoupling. Accordingly, constitutive activation of this facilitatory pathway, in the absence of agonist, may operate (perhaps even representing a disease-specific, genetic polymorphism of the 5-HT$_4$ receptor?). Thus, 5-HT$_4$ receptor antagonists with negative efficacy (inverse agonists) could have utility regardless of endogenous 5-HT levels.

5-HT is found primarily in the brain (neurones), enterochromaffin (EC) and enterochromaffin-like cells (ECL), mast cells and platelets. A large amount of 5-HT is created in the "amine precursor uptake and decarboxylation" (APUD) cells of the gastrointestinal tract, and other peripheral tissues (including LUT).[5,38] These cells have local "neuroendocrine" or "paraneuronal" properties, and release large quantities of 5-HT into the extracellular fluid and blood circulation. In the blood circulation, 5-HT is rapidly bound by specific binding proteins and platelets, so that little exists free in the plasma.[39] The platelet-derived 5-HT represents a large, and highly available source of the amine.

If one draws parallels from the alimentary tract, where the tonic role of 5-HT in the control of peristalsis and fluid secretion is better established, the sources of 5-HT—mainly in the mucosa and in significant amounts in the myenteric plexus—have been more clearly identified.[40,41] In the gastrointestinal wall, 5-HT-containing enteric interneurones have been described which terminate in the submucous ganglia,[42] and play an important modulatory role. EC and ECL cells are distributed widely along the gastrointestinal submucosa, and are the major 5-HT-containing cells of the diffuse neuroendocrine or APUD system. Similar cells have been described to be densely localized in prostate, urethra and to some extent in bladder mucosa;[43] however, they are only sparsely present in bladder body. In the gut, these cells are involved in mechanical and chemical sensation, and respond to changes in local environmental factors, such as intraluminal pressure, during peristalsis, and secrete 5-HT which then coordinates and fine-tunes the intrinsic and extrinsic nervous reflexes in the control of motility.[44] As described by Iwanaga et al[44] "...neuro-paraneuronal connection in the urethra may play an important role in the serotonin-evoked urethrogenital reflex. Intestinal and urethral serotonin-containing paraneurons, which are sensory in nature, may release serotonin in response to luminal stimuli, and directly activate adjacent peptidergic neurons to initiate the reflex arcs." In sheep, dogs and man, bladder, urethral and prostatic epithelia contain a large number of amine- and/or peptide-producing neuroendocrine cells (also called paraneurons) which are rich in 5-HT.[45-47] It is supposed that the presence of 5-HT in the urogenital tract is in part functionally correlated with the emission of urine and semen. Whether a population of such cells performs similar functions throughout the bladder, fine-tuning the interplay of storage and voiding, has not been demonstrated.

A second source of 5-HT in the bladder is the mast cell.[48,49] Mast cells containing 5-HT positive granules are present in the urothelium and submucosa of the bladder. In extreme sensory disorders of the bladder, and in particular, in interstitial cell cystitis (ICC), activated mast cells have been shown clearly to be distributed throughout the bladder wall, even into the smooth muscle layers.[49] Thus, the likelihood that mast cell secreted 5-HT can modulate sensory and motor function in the bladder of ICC patients is significant.

Indeed, it has been demonstrated that mast cells can release 5-HT in response to muscarinic receptor stimulation.[50] In this study, carbachol triggered rat bladder mast cell serotonin release in a dose-dependent manner, an effect increased by tissue pretreatment with estradiol and blocked by atropine. The effect of carbachol was accompanied by ultrastructural evidence of mast cell activation and was stronger than that obtained by either compound 48/80 or substance P. The conclusion, that bladder mast cell activation is neurogenically mediated and augmented by estradiol, could possibly explain not only the painful symptoms of ICC and its prevalence in women, but also the worsening of symptoms perimenstrually. ICC is an extreme example of disordered bladder function, associated with probably many, hitherto unidentified factors, but it represents a relatively "minor" pharmaceutical market, despite the clear unmet therapeutic need. However, it has also been suggested that women without clear ICC (e.g., refractory idiopathic instability and idiopathic sensory urgency) may have certain features in common, such as activated mast cell migration into the detrusor.[51,52] If this is true, and thus provides a real source of endogenous 5-HT in the bladder, then the mechanistic basis and potential patient population for a selective 5-HT₄ receptor antagonist gains considerable significance.

Frequently listed as a bountiful source of 5-HT in the body is the platelet.[53] Many physiological activities associated with 5-HT receptor activation and even the basis for certain pathophysiology, have been proposed to derive from or be associated with platelet 5-HT (e.g., cerebrovascular dilatation in migraine;[54] atrial arrhythmias[55]). No evidence links platelet 5-HT content to bladder activity, although it remains feasible that altered vascular reactivity in visceral organs such as bladder could cause platelet shape change and release.

Examples of 5-HT$_4$ Receptor Function in LUT from Monkey and Man

Methods

The following methods were utilized in these studies of bladder tissues from monkey,[14] and man.[17] Rectangular strips (2 cm x 0.5 cm) of urinary bladder from *Rhesus* and *Cynomolgus* monkeys were taken from the posterior medial smooth muscle layer and mounted vertically between two platinum wire electrodes in organ baths containing Tyrode solution. Specimens from the anterior dome of the human urinary bladder were obtained from patients undergoing cystectomy en bloc due to bladder malignancy, or from bladder augmentation procedures performed on spina bifida patients. Immediately after surgical removal, the urothelium and serosal layers of tissue were dissected away and muscular strips (10 mm long x 3 mm wide x 3 mm thick) were cut and mounted vertically between platinum wire electrodes and placed in organ baths containing modified Krebs solution. All tissue were kept at 37°C, pH 7.4 and gassed with O_2/CO_2, (95:5%). Cocaine (30 µM), corticosterone (30 µM), methysergide (1 µM), ondansetron (1 µM) and indomethacin (1 µM) were added to the physiological solutions in order to achieve equilibrium conditions and, importantly, to isolate pharmacologically the putative 5-HT$_4$ receptor for study.

Tissues were placed under an initial tension of 10 mN (monkey), or 20 mN (human) and subjected to electrical field stimulation (trains of electrical pulses of 3-5 s duration, supramaximal voltage, 0.1-1 ms pulse width, 20 Hz, 1 per min), yielding reproducible contractile responses. The effects of tetrodotoxin (TTX) and atropine on the responses to field stimulation were measured. Following stabilization of electrically-evoked contractions, cumulative concentration-inhibition (E/[A]) curves to agonist were constructed. Upon washout of agonist and subsequent recovery of the response to field stimulation, tissues were equilibrated with antagonist and second E/[A] curves were constructed (in some tissues, a single-curve paradigm was adopted).

To elicit direct stimulation of muscle cells, tissues were subjected to 15-30 min of electrical stimulation followed by the administration of TTX and atropine so as to inhibit completely the neurogenic contractile response. Subsequently, the pulse width of the electrical stimulation was increased 10-fold (to 10 ms) in order to evoke reproducible contractile responses of muscle similar in magnitude to those generated neurogenically. Segments of urinary bladder not subjected to electrical stimulation, were contracted with carbachol. Following the development of a sustained contracture, attempts were made to relax the tissue with 5-HT. Antagonists used in this study included DAU 6285,[23] GR 113808[20] and RS-100235.[56]

Results

Transmural electrical field stimulation evoked reproducible contractile responses in strips of monkey urinary bladder body (Fig. 8.2a) and human urinary bladder body (Fig. 8.3) and base (Fig. 8.4). These responses were blocked by TTX (3 μM) and atropine (1 μM). In the presence of TTX and atropine, broader pulse width stimulation elicited contractions by direct depolarization of the smooth muscle cells (Fig. 8.2b). These contractions were also resistant to lignocaine (100 μM) and ω-conotoxin GVIA (0.1 μM), militating against the possibility of TTX and atropine-resistant neuronal transmission in this tissue.

Figure 8.2a shows that cumulative addition of 5-HT (1 nM - 1 μM) inhibited field-stimulated contractile responses in monkey bladder in a concentration-dependent manner. 5-Methoxytryptamine and renzapride acted as full agonists, while the selective 5-HT$_{1A}$ receptor agonist, 8-OH-DPAT was without effect (data not shown). In the presence of the marginally selective 5-HT$_4$ receptor antagonist, DAU 6285 (10 μM; Fig. 8.2a), higher concentrations of 5-HT were required, causing parallel dextral displacements of inhibition curves to 5-HT. In Figure 8.2b it can be seen that similar concentrations of 5-HT were able to inhibit the responses to direct smooth muscle stimulation. The 5-HT$_4$ receptor antagonist, GR 113808, at a high concentration (3 μM) reversed this inhibition (upper trace), which was sustained in the absence of antagonist (lower trace). Electrical stimulation of strips taken from other regions of the monkey bladder (dome and bladder neck) evoked contractile responses of consistent magnitude which were also inhibited by 5-HT (data not shown), confirming that the inhibitory response to 5-HT in monkey bladder is not limited by anatomical region.

In direct contrast to the effects seen in monkey bladder, 5-HT concentration-dependently enhanced neurogenic contractile responses to field stimulation in human bladder strips. This was observed in the bladder body (see Fig. 8.3) and bladder base (see

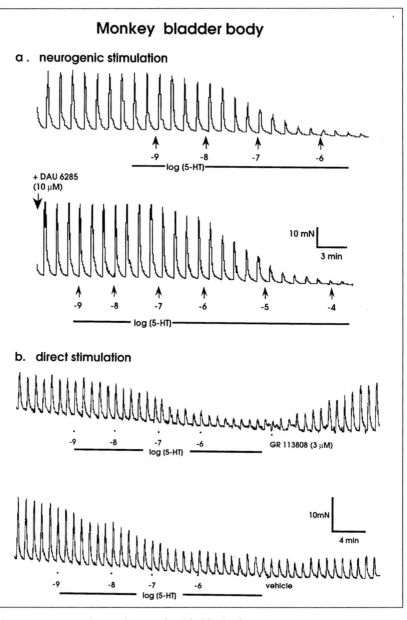

Fig. 8.2. Responses observed in monkey bladder body.

     a. Attenuation of contractile responses to *neurogenic* electrical field stimulation (20 Hz for 5 s, 1 ms pulse width, every 1 min, supramaximal voltage) by 5-HT alone (1 nM - 1 μM; upper trace), and by higher concentrations of 5-HT in the presence of DAU 6285 (10 μM; lower trace).

     b. Increasing pulse width to 10 ms (5 s duration, supramaximal voltage, 20 Hz, 1 train per min) in the presence of tetrodotoxin (3 μM) and atropine (1 μM) evoked contractions via *direct* stimulation of the muscle which were sensitive to inhibition by

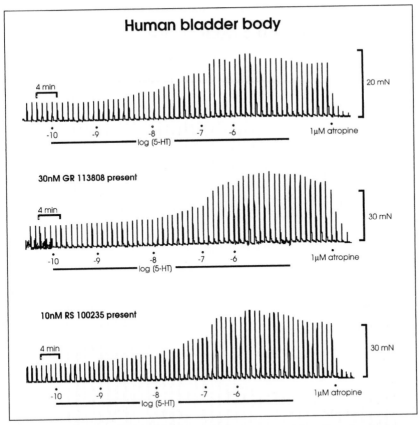

Fig. 8.3. Responses observed in human bladder body. Potentiation of responses to *neurogenic* electrical field stimulation (20 Hz for 5 s, 1 ms pulse width, every 1 min, supramaximal voltage) by 5-HT (0.1 nM - 1 µM). Potentiation by 5-HT was blocked by the selective 5-HT$_4$ receptor antagonists GR 113808 (30 nM; pA$_2$ = 8.1; middle trace) and RS-100235 (10 nM; bottom trace; pA$_2$ = 8.6). All contractions were greatly inhibited by atropine (1 µM), indicating that release of ACh and action at muscarinic cholinoceptors mediated these contractile responses.

5-HT (1 nM - 1 µM). This inhibitory effect of 5-HT was readily reversed by the novel, selective 5-HT$_4$ receptor antagonist GR 113808 (3 µM; upper trace) whereas the addition of vehicle to a corresponding strip (lower trace) did not affect the inhibition caused by 5-HT. Similar inhibition was observed against the sustained contraction evoked by carbachol (not shown).

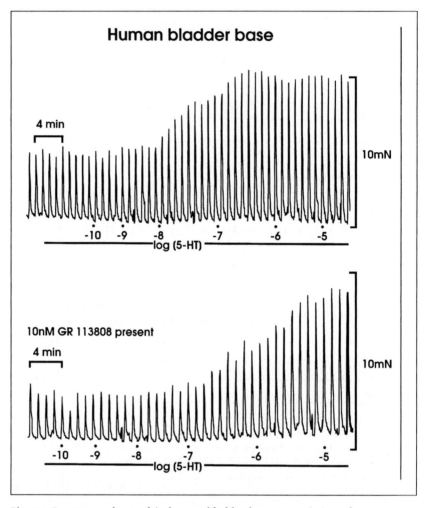

Fig. 8.4. Responses observed in human bladder base. Potentiation of responses to *neurogenic* electrical field stimulation (20 Hz for 5 s, 1 ms pulse width, every 1 min, supramaximal voltage) by 5-HT (0.1 nM - 10 µM). Facilitatory action of 5-HT was blocked by the 5-HT$_4$ receptor antagonist GR 113808 (10 µM; pA$_2$ = 9.6; lower trace).

Fig. 8.4), and surmountable antagonism by low concentrations of GR 113808 and RS-100235 confirmed that a population of 5-HT$_4$ receptors mediated the response to 5-HT. As can be seen in these figures, the facilitatory action of 5-HT was very potent and of considerable magnitude. In some tissues, facilitation occurred with 5-HT concentrations as low as 3 pM, and maximal 5-HT$_4$ receptor mediated enhancement varied between 100 and 300% of control responses.

In most tissues, responses to field stimulation were sensitive to atropine (1 μM), with residual responses eliminated by α, β-methylene-ATP treatment.

Discussion

The data exemplified above, in addition to those referred to in the introduction, demonstrate that 5-HT augments electrically-evoked contractions of isolated urinary bladder of man through activation of a $5\text{-}HT_4$ receptor. This excitatory action of 5-HT via $5\text{-}HT_4$ receptors contrasts starkly with its inhibitory action in strips of isolated bladder from *Rhesus* and *Cynomolgus* monkeys. The mechanism believed to operate in the bladder of man, which is depicted schematically in Figure 8.1, would appear to be unique to man. Only in bladder of guinea pig has any tentative evidence been reported for similar $5\text{-}HT_4$ receptor function,[25] and this requires confirmation. The uniqueness of functional localization to human bladder eliminates the possibility that an animal model may be developed to investigate further the physiological relevance of this mechanism. It is for this reason that the therapeutic potential awaiting selective $5\text{-}HT_4$ receptor antagonists for various LUT disorders remains speculative until clinical studies have been undertaken. More generally, such species-dependent variation, which is seen also with respect to localization of $5\text{-}HT_4$ receptors in atrial and adrenal tissues (see relevant chapters within this volume), illustrates clearly the ultimate importance of investigating human tissues, diseased and normal, in the quest for new therapeutic avenues. The significance of this factor is only likely to develop more with the progressive discovery of greater receptor subtype diversity.

It is obvious that the true potential of $5\text{-}HT_4$ receptor antagonists in the treatment of LUT disorders rests tremendously on the assumption that endogenous 5-HT plays an endogenous tonic modulatory role in bladder function, at least in disease states. The same holds true for potential therapy in disorders of the gastrointestinal tract and heart, although in the former, a greater body of evidence already supports such a role. The sensitivity of tissues to 5-HT is a critical issue. Our studies and those published have emphasized the low concentrations of 5-HT required to influence bladder detrusor function; clearly, a detailed investigation of how sensitivity to $5\text{-}HT_4$ receptor facilitation changes with underlying pathology would be valuable.

In light of the long-standing role of 5-HT in neuromodulation,[18,42,55,57] it is now acknowledged that the 5-HT$_4$ receptor can play a key physiological role in modulating neuromuscular activity, particularly with regard to the regulation of movement and coordinated rhythmic activity in mammalian hollow organs (alimentary tract, urinary bladder, atria of heart). Although the absolute source of 5-HT for such a role is not readily apparent (at least in heart and urinary bladder), its examination will undoubtedly be possible soon.

It is postulated, therefore, that the 5-HT$_4$ receptor is integrated into neuromuscular circuits in LUT tissues so as to facilitate coordinated rhythmic activity of a functional nature (i.e., contraction of the urinary bladder). This control may be manifested as inhibition in some species or excitation in others. Nevertheless, it provides a tremendous opportunity for the development of therapeutic agents for rhythmic disorders such as urge urinary incontinence (detrusor instability, hyperreflexia), interstitial cystitis, detrusor hypomotility, as well as atrial arrhythmias, gastro-esophageal reflux disorders, gastroparesis and irritable bowel syndrome.

Acknowledgments

The efforts of Dr. Linda M.D. Shortliffe, Department of Urology, Stanford University Medical Center, in providing samples of bladder from surgery are gratefully acknowledged.

References

1. Delaere KP, Debruyne FM, Booij LH. Influence of ketanserin (serotonin antagonist) on bladder and urethral function. Urology 1987; 29(6):669-73.
2. Andersson KE. Pharmacology of lower urinary tract smooth muscles and penile erectile tissues. Pharmacol Rev 1993; 45:253308.
3. Ormsbee HS, Fondacaro JD. Minireview: Action of serotonin in the gastrointestinal tract. Proc Soc Exp Biol Med 1985; 178:333-338.
4. Ford APDW, Clarke DE. The 5-HT$_4$ receptor. Med Res Rev 1993; 13(6):633-662.
5. Villalon CM, de Vries P, Saxena PR. Serotonin receptors as cardiovascular targets. Drug Discovery Today, 1997; 2(7):294-300.
6. Gerald C, Adham N, Kao et al. The 5-HT$_4$ receptor: molecular cloning and pharmacological characterization of two splice variants. EMBO J 1995; 14:2206.
7. Cohen ML. Canine, but not rat bladder contracts to serotonin via activation of 5HT$_2$ receptors. J Urology 1990; 143(5):1037-40.
8. Callahan SM, Creed KE Electrical and mechanical activity of the isolated lower urinary tract of the guinea-pig. Br J Pharmacol 1981; 74:353-358.

9. Saxena PR, Heiligers J, Mylecharane EJ et al. Excitatory 5- hydroxytryptamine receptors in the cat urinary bladder are of the M- and 5-HT$_2$-type. J Auton Pharmacol 1985; 5:101-107.

10. Holt SE, Cooper M, Wyllie JH. On the nature of the receptor mediating the action of 5-hydroxytryptamine in potentiating responses of the mouse urinary bladder strip to electrical stimulation. Naunyn-Schmiedeberg's Arch Pharmacol 1986; 334(4):333-40.

11. Cleal A, Corsi M, Feniuk W et al. Potentiating action of 5-hydroxytryptamine (5-HT) on electrically-induced contractions in the mouse urinary bladder. Br J Pharmacol 1989; 97:564P.

12. Hills J, Meldrum LA, Klarskov P et al. A novel non-adrenergic, noncholinergic nerve-mediated relaxation of the pig bladder neck: an examination of possible neurotransmitter candidates. Eur J Pharmacol 1984; 99:287-293.

13. Bowers CW, Kolton L. The efferent role of sensory axons in nerve-evoked contractions of bullfrog bladder. Neurosci 1987; 23:1157-1168.

14. Waikar MV, Ford APDW, Clarke DE. Evidence for an inhibitory 5-HT$_4$ receptor in urinary bladder of *Rhesus* and *Cynomolgus* monkeys. Br J Pharmacol 1994; 111(1):213-218.

15. Klarskov P,Hørby-Petersen J. Influence of serotonin on lower urinary tract smooth muscle in vitro. Br J Urol 1986; 58:507-513.

16. Hindmarsh JR, Idowu OA, Yeates WK et al. Pharmacology of electrically evoked contractions of human bladder. Br J Pharmacol 1977; 61:115P.

17. Corsi M, Pietra C, Toson G et al. Pharmacological analysis of 5-hydroxytryptamine effects on electrically stimulated human isolated urinary bladder. Br J Pharmacol 1991; 104(3):719-25.

18. Bockaert J, Fozard JR, Dumuis A et al. The 5-HT$_4$ receptor: a place in the sun. Trends in Pharmacol Sci 1992; 13:141-145.

19. Tonini M, Messori E, Franceschetti GP et al. Characterization of the 5-HT receptor potentiating neuromuscular cholinergic transmission in strips of human isolated detrusor muscle. Br J Pharmacol 1994; 113(1):1-2.

20. Gale JD, Grossman CJ, Whitehead JWF et al. GR 113808: a novel selective antagonist with high affinity at the 5-HT$_4$ receptor. Br J Pharmacol 1994; 111,332-338.

21. Candura SM, Messori E, Franceschetti GP et al. Neural 5-HT$_4$ receptors in the human isolated detrusor muscle: effects of indole, benzimidazolone and substituted benzamide agonists and antagonists. Br J Pharmacol 1996; 118(8):1965-70,.

22. Gale JD, Reeves JJ, Bunce KT. Antagonism of the gastroprokinetic effect of metoclopramide by GR 125487, a potent and selective 5-HT$_4$ receptor antagonist. J Gastrointest Motil 1993; 5:192.

23. Schiavone A, Giraldo E, Giudici L et al. DAU 6285: a novel antagonist at the putative 5-HT$_4$ receptors. Life Sci 1992; 51:583-592.

24. Eglen RM, Bley K, Bonhaus DW et al. RS-23597-190: a potent and selective 5-HT$_4$ receptor antagonist Br J Pharmacol 1993; 115: 1387-1389.

25. Messori E, Rizzi CA, Candura SM et al. 5-Hydroxytryptamine receptors that facilitate excitatory neuromuscular transmission in the guinea-pig isolated detrusor muscle. Br J Pharmacol 1995; 115(4): 677-683.

26. McCallum RW, Prakash C, Campoli-Richards D et al. Cisapride: A preliminary review of its pharmacodynamic and pharmacokinetic properties, and therapeutic use as a prokinetic agent in gastrointestinal motility disorders. Drugs 1988; 36:652-681.

27. McCallum RW. Cisapride: a new class of prokinetic agent. Am J Gastroenterol 1991; 86:135-149.

28. Nestler JE, Stratton MA, Hakim CA. Effect of metoclopramide on diabetic neurogenic bladder. Clin Pharm 1983; 2:83-85.

29. Kumar BB. Urinary incontinence associated with metoclopramide. JAMA 1984; 251 (12):1553.

30. Binnie NR, Creasey GH, Edmond P et al. The action of cisapride on the chronic constipation of paraplegia. Paraplegia 1988; 26:151-158.

31. Boyd IW, Rohan AP. Urinary disorders associated with cisapride. Med J Aus 1994; 160:579-580.

32. Pillans PI, Wood SM. Cisapride increases micturition frequency. J Clin Gastroenterol 1994; 19(4):336-346.

33. Franceschetti GP, Candura SM, Vicini D et al. Cisapride enhances detrusor contractility and improves micturition in a woman with lazy bladder. Scand J Urol Nephrol 1997; 31(2):209-210.

34. Tonini M, Candura SM. 5-HT$_4$ Receptor agonists and bladder disorders. Trends Pharmacol Sci 1996; 17(9):314.

35. Vaidyanathan S, Rao MS, Chary KS et al. Clinical import of serotonin activity in the bladder and urethra. J Urol 1981; 125(1):42-3.

36. Physicians Desk Reference, 51st edition. Montvale, NJ: Product Information. Medical Economic Company, 1997; 935:2828.

37. Sancho A, Martinez-Mir I, Palop V. Pollakiuria and micturition syndrome related to fluoxetine (letter). Medicina Clinica 1995; 105(15): 598-599.

38. Houston D, van Houtte P. Serotonin and the vascular system: role in health and disease and implications for therapy. Am J Hypertens 1988; 1:317S-328S.

39. Hollenberg N. Serotonin and the peripheral circulation. J Hypertens 1986; 4(suppl 1):S23-S27.

40. Nemoto N, Kawaoi A, Shikata T. Immunohistochemical demonstration of serotonin (5-HT) containing cells in the human and rat small intestine. Biomed Res 1982; 3:181-187.

41. Petterson G, Ahlman H, Dahlstrom A et al. The effect of transmural field stimulation on the serotonin content in rat duodenal enterochromaffin cells—in vitro. Acta Physiol Scand 1979; 107:83-87.

42. Furness JB, Costa M. Neurones with 5-hydroxytryptamine-like immunoreactivity in the enteric nervous system: Their projections in the guinea-pig small intestine. Neuroscience 1982; 7:341-349.

43. Fetissof F, Dubois MP, Arbeille-Brassart B et al. Endocrine cells in the prostate gland, urothelium and Brenner tumors. Immunohisto-

logical and ultrastructural studies. Virchow Arch B, Cell Pathology and Molecular Pathology 1983; 42(1):53-64.

44. Iwanaga T, Han H, Hoshi O et al.Topographical relation between serotonin-containing paraneurons and peptidergic neurons in the intestine and urethra. Biological Signals 1994; 3(5):259-70.

45. Iwanaga T, Hanyu S, Fujita T. Serotonin-immunoreactive cells of peculiar shape in the urethral epithelium of the human penis. Cell Tiss Res 1987; 249(1):51-6.

46. Hanyu S, Iwanaga T, Kano K et al. Distribution of serotonin-immunoreactive paraneurons in the lower urinary tract of dogs. Am J Anatomy 1987; 180(4):349-56.

47. Vittoria A, La Mura E, Cocca T et al. Serotonin-, somatostatin- and chromogranin A-containing cells of the urethro-prostatic complex in the sheep. An immunocytochemical and immunofluorescent study. J Anatomy 1990; 171:169-78.

48. Christmas TJ, Rode J Characteristics of mast cells in normal bladder, bacterial cystitis and interstitial cystitis. Br J Urol 1991; 68(5): 473-8.

49. Letourneau R, Sant GR, el-Mansoury M et al. Activation of bladder mast cells in interstitial cystitis. Int J Tissue React 1992; 14(6):307-12.

50. Spanos C, Elmansoury M, Letournea R et al. Carbachol-induced bladder mast cell activation—augmentation by estradiol and implications for interstitial cystitis. Urology 1996; 48(5):809-816.

51. Moore KH, Nickson P, Richmond DH et al. Detrusor mast cells in refractory idiopathic instability. Br J Urol 1992; 70(1):17-21.

52. Frazer MI, Haylen B, Sissons M. Do women with idiopathic sensory urgency have early interstitial cystitis? Br J Urol 1990; 66(3):274-8.

53. Frishman WH, Huberfeld SH, Okin S et al. Serotonin and serotonin antagonism in cardiovascular and non-cardiovascular disease. J Clin Pharmacol 1995; 35:541-572.

54. Fozard JR. The 5-hydroxytryptamine-nitric oxide connection: the key link in the initiation of migraine? Arch Int Pharmacodyn 1995; 329:111-119.

55. Kaumann AJ. Do human atrial 5-HT$_4$ receptors mediate arrhythmias? Trends Pharmacol Sci 1994; 15:451-455.

56. Clark RD, Jahangir A, Flippin LA et al. RS-100235: A high affinity 5-HT$_4$ receptor antagonist. Bioorg Med Chem Lett 1995; 5:2119-2122.

57. Hen R. Structural and functional conservation of serotonin receptors throughout evolution. In: Pichon Y, ed. Comparative Molecular Neurobiology. Basel: Birkhäuser Verlag, 1993:266-278.

CHAPTER 9

The 5-HT$_4$ Receptor in the Adrenal Gland

Hervé Lefebvre, Vincent Contesse, Catherine Delarue,
Jean-Marc Kuhn and Hubert Vaudry

Introduction

Serotonin (5-HT) plays a pivotal role in the control of various neuroendocrine functions.[1] In particular, 5-HT regulates the secretory activity of the adrenal cortex through multiple actions at different levels of the hypothalamo-pituitary adrenal (HPA) axis (for a review see ref. 2). In the central nervous system, 5-HT stimulates corticotropin-releasing hormone (CRH) released from hypothalamic neurons.[3,4] 5-HT also triggers adrenocorticotropin (ACTH) secretion from corticotrophs in the anterior pituitary.[5,6] The physiological relevance of these findings is supported by the observations that central serotonergic pathways are involved in the circadian rhythmicity of ACTH secretion[7] and in the activation of the HPA axis during stress.[8] More recently, it has been shown that the receptors mediating the stimulatory effect of 5-HT on ACTH secretion belong to the 5-HT$_{1A}$, 5-HT$_{2A/2C}$ and/or 5-HT$_3$ subtypes.[6,8] Apparently, 5-HT$_4$ receptors are not involved in the stimulatory effect of 5-HT on CRH and/or ACTH secretion.[9,10]

The renin-angiotensin system, which plays a major role in the regulation of aldosterone secretion, can also mediate the corticotropic effect of 5-HT. Clinical studies have shown an increase in plasma renin activity (PRA) following the administration of 5-HT precursors to healthy volunteers.[11] In rat, an increase in PRA has been observed after intracerebroventricular injections of 5-HT,[12] suggesting the existence of a central serotonergic control of renin secretion.

5-HT$_4$ Receptors in the Brain and Periphery, edited by Richard M. Eglen.
© 1998 Springer-Verlag and R.G. Landes Company.

Pharmacological studies have shown that the effect of 5-HT on renin secretion involves the activation of 5-HT$_{1B}$, 5-HT$_{2A}$ and 5-HT$_{2C}$ receptor subtypes.[12]

Beside its indirect effects via ACTH and renin secretion, 5-HT is also capable of directly influencing corticosteroid production from adrenocortical cells via a paracrine mechanism. The occurrence of 5-HT in the adrenal gland has been demonstrated by immunocytochemical and biochemical techniques. In frog and rat, 5-HT is synthesized and released by adrenochromaffin cells[13-15] while in man, 5-HT is produced by intra-adrenal mast cells.[16,17] Concurrently, it has been shown that 5-HT stimulates gluco- and mineralocorticoid secretion in various species including man.[18-22] In rat, the type of serotonergic receptor mediating the action of 5-HT on adrenocortical tissue remains disputed. The involvement of a 5-HT$_2$ receptor type has been proposed on the basis of pharmacological data.[23,24] However, it has been clearly demonstrated that in adrenocortical cells, 5-HT increases cyclic adenosine monophosphate (cAMP) production[23] but does not influence phosphoinositide turnover.[25] Since the 5-HT$_2$ receptor is positively coupled to phospholipase C,[26] it appears that the effect of 5-HT on corticosteroid secretion cannot be ascribed to activation of a 5-HT$_2$ receptor subtype. In the present chapter, we summarize the biochemical and functional evidence supporting the involvement of a 5-HT$_4$ receptor subtype in the mechanism of action of serotonin on frog and human adrenocortical cells.

Pharmacological Characterization of 5-HT$_4$ Receptors in the Adrenal Gland

The initial studies aimed at characterizing the serotonergic receptor in the frog adrenal gland revealed that the stimulatory effect of 5-HT (1 µM) was not blocked by various 5-HT$_1$, 5-HT$_2$ or 5-HT$_3$ antagonists such as metergoline, methiothepine, mesulergine, methysergide, ketanserin and MDL 72222 employed at concentrations up to 10 µM.[27] In addition, the 5-HT$_1$ receptor agonist 8-hydroxydipropylaminotetraline (8-OH-DPAT) and the 5-HT$_3$ receptor agonist 2-methyl-5-HT did not mimic the stimulatory effect of 5-HT on corticosteroid production.[28] These data clearly showed that in amphibians, the corticotropic action of 5-HT cannot be accounted for by recruitment of 5-HT$_1$, 5-HT$_2$ or 5-HT$_3$ receptors. In contrast, the effect of 5-HT on corticosteroid production was completely blocked by a high concentration (500 µM) of the indoletropane derivative ICS 205930,[28] suggesting that a 5-HT$_4$ receptor could be involved in the mechanism of action of 5-HT on frog adrenocortical cells. In order to test this hypothesis we have investigated the effect of two series of 5-HT$_4$ agonists, i.e., substituted benzamide derivatives and azabicycloalkyl benzimidazolones derivatives. All 5-HT$_4$ receptor agonists studied caused a dose-dependent stimulation of corticosteroid secretion. Among the benzamide derivatives tested, BRL 24924 renzapride was the most potent compound followed by cisapride and (R,S)-zacopride, the latter being the most efficient to stimulate corticosteroid secretion.[28] The intrinsic activity of (R,S)-zacopride was also compared to that of its two enantiomers. (S)-zacopride exhibited the highest efficacy, the rank of order being: (S)-zacopride > (R,S)-zacopride > (R)-zacopride.[29] In contrast, all three compounds exhibited similar potencies. The relative efficacy of the two benzimidazolone derivatives studied, i.e., BIMU 1 and BIMU 8, on corticosteroid secretion was similar to that previously reported in mouse embryo colliculi neurons,[30] BIMU 8 being the most efficient compound.[29] We have also investigated the effects of two selective 5-HT$_4$ receptor antagonists, in addition to ICS 205930 tropisetron, on zacopride-evoked corticosteroid secretion: the benzimidazolone derivative DAU 6285 and the indoleamine derivative GR113808.[29] The order of potency of these antagonists (GR113808 > DAU 6285 > ICS 205930) was similar to that previously determined in various animal models (Fig. 9.1). Moreover, the pK$_i$ values against (R,S)-zacopride (6.20 for ICS 205930, 7.84 for DAU 6285 and 10.34 for GR 113808) were in good agreement with those reported in mouse colliculi neurons[31] or in the guinea-pig striatum and hippocampus.[32]

In order to study the pharmacological characteristics of the 5-HT receptor present in the human adrenal cortex, we have investigated the effect of various serotonergic agonists and antagonists on corticosteroid secretion from human adrenocortical slices. The adrenal glands were obtained from expanded nephrectomies for kidney cancer. In man, as previously observed in frog, methysergide and ketanserin (1 µM) did not affect the response to 5-HT, while ICS 205930 (1 µM) completely inhibited 5-HT-induced cortisol production.[16] (R,S)-zacopride caused a dose-dependent increase in both cortisol and aldosterone secretion (Fig. 9.2a,b). Interestingly, (R,S)-zacopride was about 3 times more efficient in stimulating aldosterone than cortisol production.[33] Attenuation of the response was observed during prolonged exposure of adrenocortical tissue to 5-HT (0.1 µM)

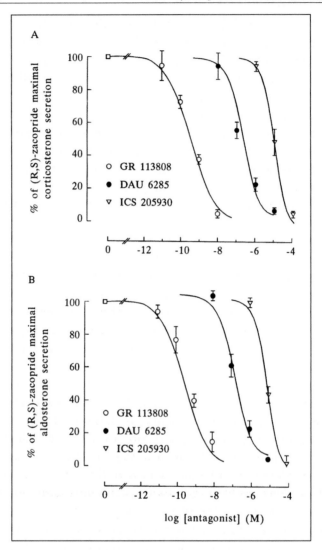

Fig. 9.1. Semilogarithmic plot comparing the effect of increasing concentrations of ICS 205930, DAU 6285 and GR 113808 on (R,S)-zacopride-induced stimulation of corticosterone (A) and aldosterone (B) secretion from perifused frog adrenal slices. Reprinted from Eur J Pharmacol 1994; 265:27-33 Contesse V et al, Effect of a series of 5-HT₄ receptor agonists and antagonists on steroid secretion by the adrenal gland in vitro, with kind permission of Elsevier Science – NL, Sara Burgerhartstraat 25, 1055 KV Amsterdam, The Netherlands.

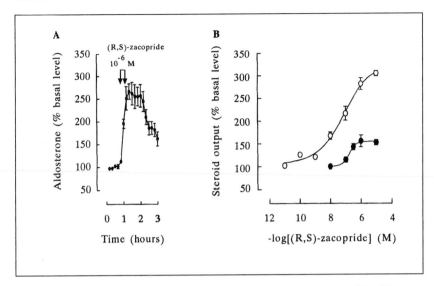

Fig. 9.2. Effect of (R,S)-zacopride on aldosterone secretion from perifused human adrenocortical slices. (A) Typical profile of the aldosterone response to a single pulse (20 min) of 10^{-6} M (R,S)-zacopride. (B) Semilogarithmic plot comparing the effects of increasing doses of (R,S)-zacopride on aldosterone (o) and cortisol (•) secretion. Reprinted with permission from Lefebvre H et al. Effect of the serotonin₄ receptor agonist zacopride on aldosterone secretion from the human adrenal cortex: in vivo and in vitro studies. J Clin Endocrinol Metab 1993; 77:1662-1666, © 1993 The Endocrine Society.

or (R,S)-zacopride (1 μM), suggesting the occurrence of a desensitization phenomenon.[16] Administration of compound 48/80 which causes degranulation of mast cells[34] provoked the release of 5-HT and a subsequent increase in aldosterone secretion from perifused human adrenocortical slices. The stimulatory effect of compound 48/80 on aldosterone production was abolished by concomitant administration of 1 μM GR 113808.[17] These pharmacological data supported the concept that, in man as in frog, the stimulatory effect of 5-HT on corticosteroid release is mediated through a 5-HT₄ receptor subtype.

Transduction Mechanisms of Adrenal 5-HT₄ Receptors

Early studies have shown that 5-HT₄ receptors are positively coupled to adenylyl cyclase.[35] In addition, further investigations have demonstrated that activation of 5-HT₄ receptors can be associated with other transduction mechanisms, such as inhibition of membrane K^+ channels in colliculi neurons[36] and stimulation of calcium current in human atrial myocytes.[37] In frog and human adrenal glands, both 5-HT (0.3 µM-3 µM) and (R,S)-zacopride (0.1 µM-10 µM) caused an increase in cAMP content.[16,28] These data provided further evidence for the involvement of an authentic $5-HT_4$ receptor in the secretory response of adrenocortical cells to 5-HT.

Using a microfluorimetric technique, we have recently shown that 5-HT induces a dose-dependent increase in cytosolic calcium concentration ($[Ca^{2+}]_i$) in cultured frog and human adrenocortical cells.[38,39] As expected, the stimulatory action of 5-HT is mimicked by (R,S)-zacopride (Fig. 9.3) and abolished by GR 113808 (0.1 µM). Several lines of evidence indicate that the increase in $[Ca^{2+}]_i$ does not result from the release of Ca^{2+} from intracellular stores, but is solely due to calcium influx through membrane channels: (1) Suppression of calcium in the culture medium by addition of 10 mM ethylene glycol bis(b-aminoethyl ether) (EGTA) completely abolished the stimulatory effect of 5-HT on $[Ca^{2+}]_i$; (2) Thapsigargin, a calcium-ATPase inhibitor that causes depletion of intracellular Ca^{2+} stores, did not affect the 5-HT-evoked $[Ca^{2+}]_i$ increase; and (3) The phospholipase C inhibitor U-73122 did not influence the rise in $[Ca^{2+}]_i$ induced by 5-HT.[38] The type of calcium channel responsible for the calcium entry associated with $5-HT_4$ receptor activation has been studied in frog adrenocortical cells. The L-type calcium channel blocker nifedipine (10 µM) and the N-type calcium channel blocker ω-conotoxin GVIA (1 µM) did not affect the calcium response. Conversely, pimozide (10 µM), a T-type calcium channel antagonist, totally blocked calcium influx induced by (R,S)-zacopride (Fig. 9.4).[38] Electrophysiological studies performed on human atrial myocytes have shown that $5-HT_4$ receptor stimulation activates a L-type calcium current.[37] These data indicate that, depending on the species or the tissue, activation of $5-HT_4$ receptors may be associated with opening of various types of voltage-operated calcium channels.

The observation that in adrenocortical cells, $5-HT_4$ receptor agonists induce both an increase in cAMP formation and a rise in $[Ca^{2+}]_i$ through T-type calcium channels, raises the question of the sequence of intracellular events involved in the mechanism of action of 5-HT. Previous studies have shown that in human atrial myocytes, the stimulatory effect of 5-HT on calcium influx is mediated through activation of a cAMP-dependent protein kinase A (PKA) pathway.[37] In frog adrenal cells, administration of dibutyryl-cAMP (dbcAMP), a permeant analog of cAMP, mimicked the action of 5-HT and $5-HT_4$ receptor agonists on $[Ca^{2+}]_i$. In addition, the phosphodi-

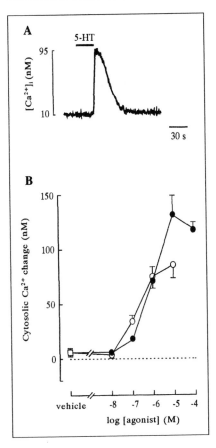

Fig. 9.3. Effect of 5-HT or zacopride on cytosolic free calcium concentration ([Ca²⁺]ᵢ) in cultured frog adrenocortical cells. (A) Typical profile illustrating the effect of a single application (30 sec) of 10^{-5} M 5-HT. (B) Semilogarithmic plot comparing the effects of increasing doses of 5-HT (o) and (R,S)-zacopride (•) on the amplitude of the [Ca²⁺]ᵢ response. Reprinted with permission from Contesse V et al, Mol Pharmacol 1996; 49:481-493. © 1996 The American Society for Pharmacology and Experimental Therapeutics.

esterase inhibitor 3-isobutyl-1-methylxanthine (IBMX) significantly increased the amplitude and the duration of the calcium response to (R,S)-zacopride. Finally, the PKA inhibitor adenosine-3',5'-cyclic monophosphorothioate (Rp-cAMPs) markedly decreased the [Ca²⁺]ᵢ rise evoked by zacopride.[38] Taken together, these data indicate that the stimulatory effect of 5-HT on [Ca²⁺]ᵢ in adrenocortical cells can also be accounted for by activation of the PKA pathway.

We next examined the respective contribution of the two transduction mechanisms, i.e., PKA and calcium influx, to the secretory response of adrenocortical cells to 5-HT. We found that the stimulatory effect of the 5-HT₄ receptor agonist zacopride on corticosteroid secretion by perifused frog adrenal slices was significantly attenuated when calcium was suppressed from the incubation medium (Fig. 9.5). Rp-cAMPs also reduced the secretory response evoked by zacopride. Finally, addition of Rp-cAMPs to a calcium-free perifusion

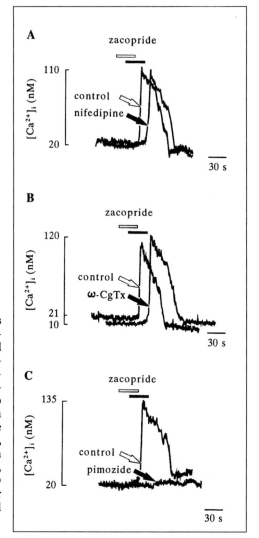

Fig. 9.4. Effects of selective blockers of voltage-sensitive calcium channels on (R,S)-zacopride-induced $[Ca^{2+}]_i$ rise in cultured adrenocortical cells. Typical profiles illustrating the effects of a single application of (R,S)-zacopride (10^{-5} M, 30 sec) in the absence (control) or in the presence of 10^{-5} M nifedipine (A), 10^{-6} M ω-conotoxin GVIA (B), or 10^{-5} M pimozide. Reprinted with permission from Contesse V et al, Mol Pharmacol 1996; 49:481-493.© 1996 The American Society for Pharmacology and Experimental Therapeutics.

medium totally abolished the effect of the 5-HT₄ receptor agonist on corticosteroidogenesis.[38] Taken together these data indicate that both the activation of PKA and the rise in $[Ca^{2+}]_i$ evoked by 5-HT₄ receptor stimulation participate to the stimulatory effect of 5-HT on corticosteroid secretion.

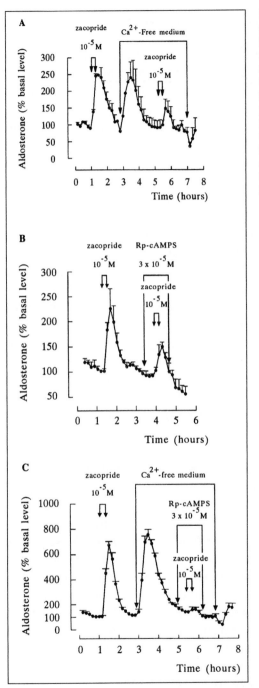

Fig. 9.5. Effects of calcium-free medium and/or the PKA inhibitor Rp-cAMPs (3×10^{-5} M) administered separately (A and B, respectively) or simultaneously (C) on the aldosterone response to (R,S)-zacopride (10^{-5} M) from perifused frog adrenal slices. Reprinted with permission from Contesse V et al, Mol Pharmacol 1996; 49:481-493. © 1996 The American Society for Pharmacology and Experimental Therapeutics.

In Vivo Effect of 5-HT$_4$ Receptor Agonists on Corticosteroid Secretion

Clinical studies aimed at investigating the effect of 5-HT$_4$ receptor agonists on corticosteroid secretion have been initially conducted in healthy volunteers. We found that a single oral dose of 400 mg (R,S)-zacopride or 10 mg cisapride (Fig. 9.6) induces a robust increase in plasma aldosterone levels, but does not affect the concentrations of potassium, renin, cortisol and ACTH.[33,40] These data indicate that the stimulatory effect of 5-HT$_4$ receptor agonists on aldosterone level cannot be ascribed to stimulation of pituitary corticotrophs, activation of the renin-angiotensin system or elevation of kalemia, but likely results from the direct action of these compounds on the adrenal cortex. The lack of effect of 5-HT$_4$ receptor agonists on cortisol level is consistent with in vitro data which showed that (R,S)-zacopride is much more efficient to stimulate aldosterone than cortisol secretion.[33]

It has long been observed that metoclopramide, the progenitor of substitutive benzamide derivatives, acutely stimulates aldosterone secretion.[41] It was generally accepted that the effect of metoclopramide on mineralocorticoid production could be accounted for by its antagonistic properties on dopamine receptors, dopamine being a potent inhibitor of glomerulosa cell activity.[42] However, this notion was not supported by the fact that other dopaminergic antagonists had no influence on plasma aldosterone level. Because metoclopramide is also a potent 5-HT$_4$ receptor agonist, it now appears that its corticotropic effect is likely attributable to stimulation of adrenal 5-HT$_4$ receptors.[43]

The 5-HT$_4$ receptor agonist cisapride is now commonly used for the treatment of gastrointestinal disorders and the doses administered are generally 3-fold higher than those required to stimulate aldosterone secretion in man. However, clinical and biological signs associated with hyperaldosteronism, i.e., hypertension and/or hypokalemia, have never been reported during prolonged administration of cisapride for gastrointestinal diseases.[44] The lack of effect of cisapride on aldosterone secretion during chronic administration of the drug is likely attributable to the desensitization phenomenon of the adrenal 5-HT$_4$ receptor previously reported in vitro.[16]

In rat, it has been shown that 5-HT, in addition to its intrinsic effect on glomerulosa cells, potentiates the stimulatory action of angiotensin II on aldosterone secretion. In contrast, 5-HT has no

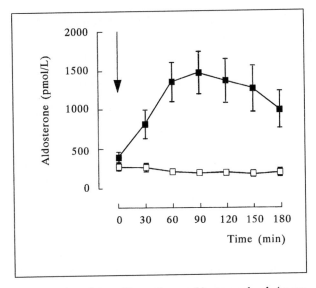

Fig. 9.6. Effect of cisapride on plasma aldosterone levels (mean ± SEM) in 8 dexamethasone-pretreated healthy volunteers. A single oral dose of 10 mg cisapride (■) or placebo (□) was administered at t_0 (arrow). Reprinted with permission from Lefebvre H, Contesse V, Delarue C et al. The serotonin-4 receptor agonist cisapride and angiotensin-II exert additive effects on aldosterone secretion in normal man. J Clin Endocrinol Metab 1995; 80:504-507; © 1995 The Endocrine Society.

effect on aldosterone secretion stimulated by ACTH.[25] We have thus explored the possible interactions between 5-HT and other corticotropic factors on aldosterone secretion in healthy volunteers. Concomitant administration of cisapride and angiotensin II or administration of cisapride after activation of the renin-angiotensin system by a low sodium diet, only induced additive effects on mineralocorticoid production,[40] indicating that in humans, 5-HT does not potentiate the stimulatory action of angiotensin II on aldosterone secretion. Combined administration of cisapride and ACTH in patients with corticotropic insufficiency revealed that cisapride does not affect ACTH-evoked aldosterone secretion.[45] Collectively, these findings are consistent with the fact that 5-HT₄ and ACTH receptors are coupled to the same transduction pathway, i.e., adenylyl cyclase,[16,35,46] while adrenal type 1 angiotensin II receptors stimulate phospholipase-C activity.[47]

It is well demonstrated that in adrenocortical cells, the expression of guanine nucleotide-binding stimulatory (G_s) protein is stimulated by prolonged administration of ACTH and angiotensin II.[48,49] In addition, ACTH and angiotensin II exert both homologous and heterologous modulation of ACTH receptors in adrenocortical cells.[50,51] It is thus conceivable that ACTH and angiotensin II could modulate the responsiveness of human adrenocortical tissue to 5-HT by regulating the expression of either the 5-HT$_4$ receptor or the stimulatory G_s protein. In order to explore this hypothesis, we have studied the effect of cisapride in patients with suppressed plasma ACTH, i.e., patients with corticotropic insufficiency (CI), and in patients with suppressed plasma angiotensin II, i.e., patients with primary hyperaldosteronism (PH). Indeed, PH is a condition characterized by an increased production of aldosterone by abnormal zona glomerulosa tissue (adenoma or bilateral adrenocortical hyperplasia).[52] Mineralocorticoid excess leads to increased sodium retention and, consequently, to an increase of the extracellular fluid volume and hypertension. The expanded plasma volume causes suppression of renin-angiotensin II activity.[52] In CI and PH patients, cisapride induced an increase in plasma aldosterone similar to that previously reported in normal volunteers,[45] indicating that prolonged suppression of ACTH and angiotensin II does not affect the responsiveness of adrenal glomerulosa cells to 5-HT$_4$ receptor agonists. The fact that cisapride was still able to stimulate aldosterone secretion in PH also shows that hyperplastic and adenomatous glomerulosa tissues express a functional 5-HT$_4$ receptor in very much the same way as normal glomerulosa cells. In agreement with this finding, 5-HT has been shown to stimulate aldosterone secretion in vitro from incubated aldosterone-producing adenoma (aldosteronoma) cells.[21]

Conclusion and Perspectives in the Treatment of Aldosterone Disorders

The data summarized in the present chapter indicate that 5-HT stimulates corticosteroid secretion in a paracrine manner, through the activation of a 5-HT$_4$ receptor subtype positively coupled to adenylyl cyclase and cytosolic calcium. The physiological relevance of this observation remains to be determined. However, it seems conceivable that intra-adrenal 5-HT may be released during stress, together with various bioactive signals, such as catecholamines and/or cytokines, and can thus directly stimulate cortisol production.

Similarly, several clinical studies indicate that aldosterone secretion is largely independent of the renin-angiotensin II system, suggesting that the secretory activity of glomerulosa cells may be physiologically regulated by intra-adrenal factors such as 5-HT. On the other hand, 5-HT may play a significant role in the physiopathology of various aldosterone disorders. In particular, the pathogenesis of PH may involve a serotonergic mechanism, as suggested by the fact that functional adrenal $5-HT_4$ receptors are present in the adrenal cortex of patients with PH. In addition, in these patients the nonspecific 5-HT antagonist cyproheptadine decreases aldosteronemia,[53] while oral administration of the serotonin precursor 5-hydroxytryptophan induces a significant increase in plasma aldosterone level.[21] Finally, the demonstration of the presence of numerous mast cells in aldosteronoma tissues[54] suggests that 5-HT, released by intratumoral mast cells, may play a significant role in the pathophysiology of these tumors by exerting tonic paracrine stimulation on aldosterone secretion. Idiopathic hypertension associated with low plasma renin is another situation in which aldosterone secretion appears to be inappropriate and autonomous, suggesting the involvement of an unknown aldosterone secretagogue.[55] It is conceivable that overproduction of adrenal 5-HT could be responsible for abnormal aldosterone secretion and hypertension. Edematous states of cirrhosis, nephrosis and heart failure are frequently associated with secondary (angiotensin II-induced) hyperaldosteronism. Activation of the renin-angiotensin system in patients with these diseases occurs in response to hypovolemia due to redistribution of volume into extravascular tissues. Although it represents a normal compensatory mechanism, the hyperaldosteronism leads to increased edema. Surprisingly, the use of angiotensin-converting enzyme inhibitors in these patients does not allow to normalize plasma aldosterone levels,[56] indicating that regulatory factors other than the renin-angiotensin II system may be involved in the pathogeny of secondary hyperaldosteronism in edematous states. Since 5-HT appears to be involved in the regulation of aldosterone secretion, it is conceivable that 5-HT can participate in the pathogeny of aldosterone overproduction in these situations.

In vivo treatment with potent and selective $5-HT_4$ receptor antagonists is urgently needed to evaluate the possible involvement of 5-HT in the physiological regulation of aldosterone secretion as well as in pathological conditions. Concurrently, development of these compounds will open new perspectives for the treatment of various aldosterone disorders such as primary hyperaldosteronism and edematous state-associated secondary hyperaldosteronism. In addition, it is now well demonstrated that plasma aldosterone levels are positively correlated with left ventricular hypertrophy in patients with essential hypertension.[57] As angiotensin-converting enzyme inhibitors are not capable of decreasing aldosteronemia in these patients,[58] novel potential inhibitors of aldosterone secretion, such as 5-HT$_4$ receptor antagonists, may be of great value for the prevention of hypertensive cardiopathy.

References

1. Frazer A, Hensler JG. Serotonin. In: Siegel GJ, Agranoff BW, Wayne-Albers R, Molinoff PB, eds. Basic Neurochemistry: Molecular, Cellular and Medical Aspects, 5th ed. New York: Raven Press, 1994:283-308.
2. Lefebvre H, Contesse V, Delarue C et al. Serotonergic control of corticosteroid secretion. J Serot Res 1994; 3:189-198.
3. Holmes MC, Di Renzo G, Beckford U et al. Role of serotonin in the control of secretion of corticotrophin-releasing factor. J Endocrinol 1982; 93:151-160.
4. Calogero AE, Bernardini R, Margioris AN et al. Effects of serotonergic agonists and antagonists on corticotropin-releasing hormone secretion by explanted rat hypothalami. Peptides 1989; 10:189-200.
5. Spinedi E, Negro-Vilar A. Serotonin and adrenocorticotropin (ACTH) release: direct effects at the anterior pituitary level and potentiation of arginine vasopressin-induced ACTH release. Endocrinology 1983; 112:1217-1223.
6. Calogero AE, Bagdy G, Burrello N et al. Role for serotonin$_3$ receptors in the control of adrenocorticotropic hormone release from rat pituitary cell cultures. Eur J Endocrinol 1995; 133:251-254.
7. Szafarczyk A, Alonso G, Ixart G et al. Serotonergic system and circadian rhythms of ACTH and corticosterone in rats. Am J Physiol 1980; 249:E219-E226.
8. Saphier D, Farrar GE, Welch JE. Differential inhibition of stressed-induced adrenocortical responses by 5-HT$_{1A}$ agonists and by 5-HT$_2$ and 5-HT$_3$ antagonists. Psychoneuroendocrinology 1995; 20:239-257.
9. Bazzani C, Fiore L, Ferrante F et al. Serotonin is involved in the ACTH-induced reversal of hemorrhagic shock in anesthetized rats. Pharmacology 1996; 52:207-215.
10. Lefebvre H, Contesse V, Delarue C et al. Lack of effect of the serotonin$_4$ receptor agonist zacopride on ACTH secretion in normal men. Eur J Clin Pharmacol 1996; 51:49-51.
11. Modlinger RS, Schonmuller JM, Arora SP. Stimulation of aldosterone, renin, and cortisol by tryptophan. J Clin Endocrinol Metab 1979; 48:599-603.

12. Rittenhouse PA, Bakkum EA, Levy AD et al. Central stimulation of renin secretion through serotonergic, noncardiovascular mechanisms. Neuroendocrinology 1994; 60:205-214.

13. Verhofstad AAJ, Jonsson G. Immunohistochemical and neurochemical evidence for the presence of serotonin in the adrenal medulla of the rat. Neuroscience 1983; 10:1443-1453.

14. Holzwarth MA, Brownfield MS. Serotonin coexists with epinephrine in rat adrenal medulla. Neuroendocrinology 1985; 41:230-236.

15. Delarue C, Becquet D, Idres S et al. Serotonin synthesis in adrenochromaffin cells. Neuroscience 1992; 46:495-500.

16. Lefebvre H, Contesse V, Delarue C et al. Serotonin-induced stimulation of cortisol secretion from human adrenocortical tissue is mediated through activation of a serotonin$_4$ receptor subtype. Neuroscience 1992; 47:999-1007.

17. Lefebvre H, Compagnon P, Contesse V et al. Production and metabolism of serotonin (5-HT) by the human adrenal cortex. Endocr Res 1996; 22:851-853.

18. Müller J, Ziegler WH. Stimulation of aldosterone biosynthesis in vitro by serotonin. Acta Endocrinol 1968; 59:23-35.

19. Racz K, Wolf I, Kiss R et al. Corticosteroidogenesis by isolated human adrenal cells: Effect of serotonin and serotonin antagonists. Experientia 1979; 35:1532-1534.

20. Al-Dujaili EAS, Boscaro M, Edwards CRW. An in vitro stimulatory effect of indoleamines on aldosterone biosynthesis in the rat. J Steroid Biochem 1982; 17:351-355.

21. Shenker Y, Gross MD, Grekin RJ. Central serotonergic stimulation of aldosterone secretion. J Clin Invest 1985; 76:1485-1490.

22. Delarue C, Lefebvre H, Idres S et al. Serotonin stimulates corticosteroid secretion by frog adrenocortical tissue in vitro. J Steroid Biochem 1988; 29:519-525.

23. Matsuoka H, Ishii M, Goto A et al. Role of serotonin type 2 receptors in regulation of aldosterone production. Am J Physiol 1985; 249:E234-E238.

24. Rocco S, Boscaro M, D'Agostino D et al. Effect of ketanserin on the in vitro regulation of aldosterone. J Hypertension 1986; 4:S51-S54.

25. Rocco S, Ambroz C, Aguilera G. Interaction between serotonin and other regulators of aldosterone secretion in rat adrenal glomerulosa cells. Endocrinology 1990; 127:3103-3110.

26. Conn PJ, Sanders-Busch E, Hoffmann B et al. A unique serotonin receptor in choroid plexus is linked to phosphatidylinositol turnover. Proc Natl Acad Sci USA 1986; 83:4086-4088.

27. Idres S, Delarue C, Lefebvre H et al. Mechanism of action of serotonin on frog adrenal cortex. J Steroid Biochem 1989; 34:547-550.

28. Idres S, Delarue C, Lefebvre H et al. Benzamide derivatives provide evidence for the involvement of a 5-HT$_4$ receptor type in the mechanism of action of serotonin in frog adrenocortical cells. Mol Brain Res 1991; 10:251-258.

29. Contesse V, Hamel C, Delarue C et al. Effect of a series of 5-HT$_4$ receptor agonists and antagonists on steroid secretion by the adrenal gland in vitro. Eur J Pharmacol 1994; 265:27-33.

30. Dumuis A, Sebben M, Monferini E et al. Azabicycloalkyl benzimidazolone derivatives as a novel class of potent agonists at the 5-HT$_4$ receptor positively coupled to adenylate cyclase in brain. Naunyn-Schmiedeberg's Arch Pharmacol 1991; 343:245-251.

31. Dumuis A, Gozlan H, Sebben M et al. Characterization of a novel 5-HT$_4$ receptor antagonist of the azabicycloalkyl benzimidazolone class: DAU 6285. Naunyn-Schmiedeberg's Arach Pharmacol 1992; 354:264-269.

32. Grossman CJ, Kilpatrick GJ, Bunce KT. Development of a radioligand binding assay for 5-HT$_4$ receptors in guinea-pig and rat brain. Br J Pharmacol 1993; 109:618-624.

33. Lefebvre H, Contesse V, Delarue C et al. Effect of the serotonin-4 agonist zacopride on aldosterone secretion from the human adrenal cortex. In vivo and in vitro studies. J Clin Endocrinol Metab 1993; 77:1662-1666.

34. Irman-Florjanc T, Erjavec F. Compound 48/80 and substance P induce release of histamine and serotonin from rat peritoneal mast cells. Agents and Actions 1983; 13:138-141.

35. Dumuis A, Bouhelal R, Sebben M et al. A non-classical 5-hydroxytryptamine receptor positively coupled with adenylate cyclase in the central nervous system. Mol Pharmacol 1988; 34:880-887.

36. Fagni L, Dumuis A, Sebben M et al. The 5-HT$_4$ receptor subtype inhibits K$^+$ current in colliculi neurons via activation of a cyclic AMP-dependent protein kinase. Br J Pharmacol 1992; 105:973-979.

37. Ouadid H, Seguin J, Dumuis A et al. Serotonin increases calcium current in human atrial myocytes via the newly described 5-hydroxytryptamine$_4$ receptors. Mol Pharmacol 1992; 41:346-351.

38. Contesse V, Hamel C, Lefebvre H et al. Activation of 5-hydroxytryptamine$_4$ receptors causes calcium influx in adrenocortical cells: involvement of calcium in 5-hydroxytryptamine-induced steroid secretion. Mol Pharmacol 1996; 49:481-493.

39. Hamel C, Contesse V, Delarue C et al. Transduction mechanisms associated with activation of adrenal 5-HT$_4$ receptors in amphibians and humans. Ann Endocrinol 1996, 57:P092.

40. Lefebvre H, Contesse V, Delarue C et al. The serotonin-4 receptor agonist cisapride and angiotensin-II exert additive effects on aldosterone secretion in normal man. J Clin Endocrinol Metab 1995; 80:504-507.

41. Norbiato G, Bevilacqua M, Raggi U et al. Metoclopramide increases plasma aldosterone in man. J Clin Endocrinol Metab 1977; 45: 1313-1316.

42. Nussdorfer GG. Paracrine control of adrenal cortical function by medullary chromaffin cells. Pharmacol Rev 1996; 48:495-530.

43. Rizzi CA. A serotonergic mechanism for the metoclopramide-induced increase in aldosterone level? Eur J Clin Pharmacol 1994; 47:377-378.
44. Tack J, Coremans G, Janssens J. A risk-benefit assessment of cisapride in the treatment of gastrointestinal disorders. Drug Safety 1995; 12:384-392.
45. Lefebvre H, Dhib M, Godin M et al. Effect of the serotonin₄ receptor agonist cisapride on aldosterone secretion in corticotropic insufficiency and primary hyperaldosteronism. Neuroendocrinology 1997; 106:0000-0000.
46. Saez J, Morera AM, Dazord A. Mediators of the effects of ACTH on adrenal cells. Adv Cyclic Nucl Res 1981; 14:563-579.
47. Timmermans PBMWM, Wong PC, Chiu AT et al. Angiotensin II receptors and angiotensin II receptor antagonists. Pharmacol Rev 1993; 45:205-251.
48. Bégeot M, Langlois D, Spiegel AM et al. Regulation of guanine nucleotide binding regulatory proteins in cultured adrenal cells by adrenocorticotropin and angiotensin II. Endocrinology 1991; 128:3162-3168.
49. Langlois D, Saez JM, Bégeot M. Regulation of G-proteins in adrenal cortex. In: Saez JM, Brownie AC, Capponi A, Chambaz EM, Mantero F, eds. Cellular and Molecular Biology of the Adrenal Cortex. Paris: Colloque INSERM/Libbey Eurotext Ltd. Volume 222. 1992:61-74.
50. Lebrethon MC, Naville D, Bégeot M et al. Regulation of corticotropin receptor number and messenger RNA in cultured human adrenocortical cells by corticotropin and angiotensin II. J Clin Invest 1994; 93:1828-1833.
51. Mountjoy KG, Bird IM, Rainey WE et al. ACTH induces up-regulation of ACTH receptor mRNA in mouse and human adrenocortical cell lines. Mol Cell Endocrinol 1994; 99:R17-R20.
52. Weinberger MH, Grim CE, Hollifield JW et al. Primary aldosteronism: diagnosis, localization and treatment. Ann Intern Med 1979; 90:386-395.
53. Gross MD, Grekin RJ, Gniadek TC et al. Suppression of aldosterone by cyproheptadine in idiopathic aldosteronism. N Engl J Med 1981; 305:181-185.
54. Aiba M, Iri H, Suzuki H et al. Numerous mast cells in an 11-deoxycorticosterone-producing adrenocortical tumor. Arch Pathol Lab Med 1985; 109:357-360.
55. Folkow B. Physiological aspects of primary hypertension. Physiol Rev 1982; 62:347-504.
56. Fouad FM, Tarazi RC, Bravo EL et al. Long term control of congestive heart failure with captopril. Am J Cardiol 1982; 49:1489-1496.
57. Duprez D, Bauwens FR, De Buyzere M et al. Influence of arterial blood pressure and aldosterone on left ventricular hypertrophy in moderate essential hypertension. Am J Cardiol 1993; 71:17A-20A.
58. Riegger GAJ, Steilner H, Hayduk K et al. Captopril in the long term treatment of essential hypertension: changes in the renin-angiotensin aldosterone system. Am J Cardiol 1982; 49:1555-1557.

Therapeutic Applications of 5-HT$_4$ Receptor Agonists and Antagonists

Gareth J. Sanger

Introduction

Metoclopramide is used to stimulate 'upper gut' motility and prevent nausea and vomiting. These clinical benefits are attributed, respectively, to 5-HT$_4$ receptor activation[23] and to antagonism at dopamine D$_2$ and/or 5-HT$_3$ receptors.[52] However, since antagonism at dopamine D$_2$ receptors within the striatum may also cause extrapyramidal reactions and akathisia, alternative 5-HT$_4$ receptor agonists have been developed, with lower affinity for the D$_2$ receptor. The most notable is cisapride. Metoclopramide and cisapride are marketed for gastrointestinal hypomotility disorders such as gastro-esophageal reflux (heartburn; mild esophagitis), functional or nonulcer dyspepsia and gastroparesis (e.g., caused by diabetic neuropathy).[7] For both drugs, the most common side-effect attributed to 5-HT$_4$ receptor activation is the promotion of loose stools and in more severe cases, diarrhea and abdominal pain.

What is the purpose of this chapter? There does not seem to be a need to develop another 5-HT$_4$ receptor agonist and since the side-effects of cisapride are relatively benign, will a 5-HT$_4$ receptor antagonist be therapeutically unnecessary? The answer to these questions are provided by the observation that metoclopramide and cisapride are only partial agonists at the 5-HT$_4$ receptor, relative to 5-HT itself. This means that the full range of activity of endogenous 5-HT at the 5-HT$_4$ receptor is not reflected by the actions of these drugs. The need for new 5-HT$_4$ receptor agonists and for the development of a selective antagonist at the receptor is therefore re-examined in the light of this knowledge.

Intrinsic Activities of 5-HT$_4$ Receptor Agonists

Although the C-terminal intracellular region of the 5-HT$_4$ receptor exists as two different splice variants, the extracellular structure of these variants and their operational pharmacology is identical.[27] Differences in 5-HT$_4$ receptor structure cannot be easily exploited in terms of identifying novel types of 5-HT$_4$ receptor agonists or antagonists.

The 5-HT$_4$ receptor positively couples to adenylate cyclase and exerts its activity via increased intracellular cAMP, activation of protein kinase A and either closure of K$^+$ channels or the opening of voltage-sensitive Ca^{2+} channels, depending on the final functional outcome within that cell (e.g., membrane depolarization, facilitation of neurotransmitter release).[6] This process of translating the function of the 5-HT$_4$ receptor may vary in efficiency between cell or tissue type because of differences in how the cascade is magnified and/or desensitized by phosphorylation.[55] Table 10.1 shows the intrinsic activities of metoclopramide and cisapride for different isolated tissues relative to 5-HT. A striking feature is that their intrinsic activities tend to be greatest within gastrointestinal tissues, reflecting the clinical selectivity of partial 5-HT$_4$ receptor agonists for the gastrointestinal tract. The rare incidence of tachycardia with cisapride[32], for example, parallels the low intrinsic activity of this compound for the 5-HT$_4$ receptor in the human atrium, compared with the gut. The exception is the intrinsic activity of 1.4 for cisapride in mouse embryo colliculi neurones. This difference seems puzzling given the absence of any clear effect of clinically-administered cisapride on colliculi function. Perhaps it can be explained by a variation in 5-HT$_4$ receptor coupling that depends on the developmental status of the tissue. Wardle & Sanger[65] found that the sensitivity of guinea-pig isolated distal colon to 5-HT$_4$ receptor activation was greatest in younger animals.

Table 10.1. The intrinsic activities of Metoclopramide and Cisapride, relative to 5-HT (1.0), in different animal and human isolated tissues

Preparation			Intrinsic Activity	
Tissue	Species	Response	MCP	Cisapride
Distal colon[67]	Guinea-pig	contraction	0.9	0.9
Ascending colon[21]	Guinea-pig	contraction	0.3	0.2
Ileum[19]	Guinea-pig	contraction	0.6	–
Ileum[8]	Guinea-pig	EFS	1.0	0.7
Esophagus muscularis mucosa[3]	Rat	relaxation	0.8	0.8
Detrusor[10]	Human	potentiation of EFS	–	0.4
Right atrium[35]	Human	increased force of contraction	–	0.3
Sinoatrial node[35,37]	Piglet	increased beating rate	–	0.4, 0.2
Embryonic colliculi[17]	Mouse	increased cAMP	–?	1.4

Therapeutic Applications of 5-HT$_4$ Receptor Agonists

Existing Partial 5-HT$_4$ Receptor Agonists

Metoclopramide and cisapride are referred to as gastrointestinal prokinetic drugs, a term which defines their ability to enhance gastrointestinal propulsion and coordinated motility.[7] In this definition, the words 'propulsion' and 'coordinated motility' imply a particular form of stimulation in which there is a facilitation of the normal pattern of gastrointestinal motility rather than the induction of a new form of motility. Drugs which induce the latter are exemplified by the cholinomimetic bethanechol and by osmotic- or fiber-based laxatives.

5-HT$_4$ receptor agonists increase coordinated propulsive motility because they stimulate and sensitize AH- (sensory neurones) and cholinergic S-type (motor or interneurones) neurones within the myenteric plexus. This increases the sensitivity of the peristaltic reflex to induction by intraluminal pressure.[29,54] 5-HT$_4$ receptor activation also increases mucus[40] and water secretion into the intestinal lumen via neuronal and non-neuronal mechanisms, thereby increasing both the propulsive and lubricant aspects of gastrointestinal function.[54]

In addition to their common therapeutic uses (see Introduction), cisapride and metoclopramide have particular relevance in specialized areas of medicine, such as the treatment of early satiety in the elderly[41] and in palliative care, where they are used to treat paraneoplastic or HIV-associated gastroparesis (and accompanying early satiety, anorexia and nausea), drug-induced delays in gastric emptying and functional bowel obstructions caused, for example, by pancreatic cancer or drugs.[25,62] There are also an increasing number of reports on the ability of 5-HT_4 receptor agonists to promote the efficiency of the micturition reflex, sometimes a prolonged and difficult task among the aged, in which there is a significant decline in the numbers of cholinergic neurones.[60]

New Partial 5-HT₄ Receptor Agonists

The rank order of the affinity of cisapride for different receptors is $5\text{-HT}_{2A} > 5\text{-HT}_4 = \alpha_1$ adrenoceptor $> 5\text{-HT}_3 = D_2$; the affinity for the 5-HT_3 and D_2 receptors is approximately 10 times less than for the 5-HT_4 receptor.[7] There have been several attempts to identify other 5-HT_4 receptor agonists which do not have meaningful affinity for the dopamine D_2 receptor.[7,23,59] Some of these also antagonize at the 5-HT_3 or 5-HT_{2A} receptor. However, it is difficult to understand what clinical advantage these compounds offer over cisapride itself. One possibility arises from the observation that 5-HT_3 receptor antagonism may reduce the occurrence of spontaneous relaxations of the lower esophageal sphincter,[51] suggesting that a compound which combined this property with 5-HT_4 receptor activation might exert a dual action on sphincter function and hence, prove to be a superior treatment for patients with mild reflux esophagitis.[59] Conversely, 5-HT_4 receptor activation might detract from the acute anti-emetic efficacy of 5-HT_3 receptor antagonism.[54]

Full 5-HT₄ Receptor Agonists

Examples of such compounds have been described by Buchheit et al.[8] In theory, these may be useful in terms of evoking a more robust response within the gastrointestinal tract, and especially in the urinary bladder where the intrinsic activity to cisapride is low. In addition, full 5-HT_4 receptor agonists would be useful in exploring 5-HT_4 receptor function within the brain. However, it does not seem sensible to develop a full agonist at the 5-HT_4 receptor since this will maximally activate all 5-HT_4 receptors, no matter where they are located. An obvious concern will be those receptors which are located in the cardiac atria, since their activation can promote tachycardia and arrhythmia.[34]

5-HT₄ Receptor Antagonists

In normal animals or in isolated tissue preparations 5-HT₄ receptor antagonism has little or no activity. If this observation continues to be supported, it suggests that the 5-HT₄ receptor has minimal physiological function. Since 5-HT₄ receptor activation has profound effects on the functions of certain organs (e.g., gastrointestinal tract), it may be necessary to think of this receptor as a means by which a tissue defends itself against a noxious event. A similar proposal has been made for the 5-HT₃ receptor within the gut.[54]

Irritable Bowel Syndrome (IBS)

The possibility that 5-HT₃ and 5-HT₄ receptors are involved in Irritable Bowel Syndrome has previously been described.[54] 5-HT can be released within the gut by many different physiological stimuli (e.g., distension, amino acids, glucose) and by noxious or pathological events, such as ischemia,[67] rotavirus infection,[50] changes in microflora content,[63] pro-inflammatory cytokines,[22] stress, catecholamines and neuronal efferent input into the gut.[9,44,54] In IBS patients, increased numbers of 5-HT-containing enterochromaffin cells have been detected within the rectum, a finding which may relate to the increased blood plasma and urine concentrations of 5-HT metabolites detected in patients with severe symptoms of IBS, although not in others.[54]

IBS may begin as a noxious event within the gut, such as a viral infection, acute inflammation and/or stress.[11,39,47] This renders the gut hypersensitive to normal functions such as eating, digestion and defecation.[38] The mechanism of hypersensitivity is not clear and although it may begin with the liberation of substances within the gut and involve increased sensitivity to afferent nerve input into the spinal cord ('wind-up'), other mechanisms must be invoked in order to explain the development of IBS and its longevity, sometimes recorded for decades. These may include long-lasting changes within the gut itself,[1] possibly involving increased availability of 5-HT (see above) or other substances[11] and/or increases in afferent nerve density within the gut and spinal cord following neurotrophic activity during the originating noxious event. There may also be structural changes in areas of the brain that perceive gastrointestinal function.[57]

A hypersensitive spinal/central nerve system is likely to react to a normal physiological event within the gut, which it now perceives to be potentially noxious. Such reactions will include the signaling of abdominal pain, as well as the urge to defecate, diarrhea (to remove the perceived source of the discomfort) or even a reduction in gastrointestinal motility, to lessen the impact of further gut motility on the perceived area of discomfort. Such reactions will involve the release of 5-HT and other substances within the gut, since stress, catecholamines and efferent nerve inputs into the gut will induce such a change (see above). The increased liberation of 5-HT will promote gastrointestinal motility, defecation and in extreme forms, diarrhea. Such 'hyperreactive' changes operating against an already-hypersensitive background will further exacerbate the symptoms of IBS and increase the perception of abdominal discomfort and other symptoms.

In theory, both $5-HT_3$ and $5-HT_4$ receptors could be involved in the etiology of IBS. However, although $5-HT_3$ receptor antagonists can reduce the severity of diarrhea in IBS patients, this is because they prevent the normal physiological function of the $5-HT_3$ receptor in the mechanisms of defecation and not because they tackle the underlying cause of hypersensitivity; such drugs risk replacing diarrhea with constipation.[54] Furthermore, since the $5-HT_4$ receptor is well coupled to its effector mechanisms within the enteric nervous system (compared with the $5-HT_3$ receptor) and mediates sustained increases in cholinergic function, small increases in 5-HT availability within the gut will readily disturb its function. This compares with the massive release of 5-HT required to activate the $5-HT_3$ receptor and evoke emesis or other symptoms.[52] As such, it seems more reasonable to focus on the use of selective $5-HT_4$ receptor antagonists for the treatment of IBS, especially since this will reduce the ability of 5-HT to evoke hypersensitivity within the propulsive and lubricant systems of the gut without greatly affecting other gastrointestinal functions. In this way, such a treatment would also reduce the hyperreactivity component of IBS, so that the spinal and central nerve processes were no longer capable of creating the positive feedback loop generated by the increased release of 5-HT. The concept can only be tested by clinical trial. Thus, although $5-HT_4$

receptor antagonists can be demonstrated to reduce 5-HT-induced defecation,[2] diarrhea,[31] Cl$^-$ and mucus secretion[40,54] as well as visceral pain mechanisms sensitized by stress or by colorectal inflammation,[45] none of these systems are true models of IBS.

Cardiac Arrhythmia

Prophylactic treatment with a selective 5-HT$_4$ receptor antagonist may prevent cardiac arrhythmia and stroke, following 5-HT release from damaged platelets at the site of a thromboembolism.[34] This hypothesis is based on the observation that 5-HT$_4$ receptor activation can increase contractile force, tachycardia and arrhythmia in human isolated atria.

Brain Pathophysiology

The distribution of 5-HT$_4$ receptor mRNA and protein, for each splice variant, is well described for the central nervous system.[20,64] Most notable is the dense distribution of the receptor in dopamine-rich areas of the brain, including the substantia nigra and corpus striatum, certain areas of the limbic system and the hippocampus. In these areas, the receptor is located on neurones containing GABA and dynorphin or ACh, but not dopamine.[12,43] 5-HT$_4$ receptor activation may be involved in the mechanisms whereby high-potassium concentrations increase GABA release from the rat substantia nigra[68] and will increase ACh release from rat frontal cortex, but not from the striatum.[13] 5-HT$_4$ receptor activation can also increase the release of dopamine, but only via the GABA/dynorphin- and/or ACh-containing neurones.[16]

To analyze the pathophysiological significance of the brain 5-HT$_4$ receptors, Gaster and Sanger[26] simplified the above data into three functional categories. These were the mechanisms of movement (substantia nigra, corpus striatum), emotion and emotional reward mechanisms (limbic system) and cognitive processes (hippocampus, cortex). Reavill et al[48] found no significant effects of 5-HT$_4$ receptor antagonism on locomotor activity induced by amphetamine, nicotine or morphine, or on circling behavior following 6-OHDA-induced lesion of the left nigrostriatal pathway. However, these experiments do not necessarily rule out the possibility that the 5-HT$_4$ receptor could play a role in a movement disorders caused by an increased 5-HT 'tone' on the system. Tantalizing suggestions of this possibility are provided by two case-reports of cisapride-evoked aggravation of Parkinsonian tremor.[56]

Reavill et al[48] also examined for the ability of 5-HT$_4$ receptor antagonism to affect reward mechanisms using the intracranial self stimulation technique and either cocaine or a brain-penetrant 5-HT$_4$ receptor antagonist. As before, no effects of 5-HT$_4$ receptor antagonism were noted. However, this approach may need to be re-examined after a suggestion that 5-HT$_4$ receptor antagonism might reduce ethanol intake in ethanol-preferring rats.[42] Thus, it may be necessary to impose the correct pathology to detect an effect of 5-HT$_4$ receptor antagonism. Studies on rodent anxiogenic-like behaviors concluded that the 5-HT$_4$ receptor has no significant role in conflict models of behavior (Geller-Seifter test),[36] but may play a modest role in the social interaction and elevated X-maze tests.[36,58] This profile of activity is similar to that obtained using selective 5-HT$_3$ receptor antagonists. It is broadly consistent with the downregulation of sensitivity to 5-HT$_4$ receptor activation in the rat CA1 pyramidal cells following chronic administration of antidepressant drugs,[5] but conflicts with the findings of Costall and Naylor[14,15] who showed that 5-HT$_4$ receptor antagonism reduced the anxiolytic-like effects of 5-hydroxytrytophan plus ritanserin or of diazepam itself, in the mouse light/dark model of anxiety; when given alone, the antagonists had no effects on mouse behavior. Nevertheless, whatever the final interpretation of these data, it is suggested that the failure of 5-HT$_3$ receptor antagonists to exert robust anxiolytic effects in human patients indicates that the restriction of 5-HT$_4$ receptor antagonist activity to similar rodent models of anxiety will lead to similar negative findings in human patients.

5-HT$_4$ receptor activation might enhance the process of cognition,[20] perhaps by inducing long-term neuronal excitability, as described for certain central and enteric nerve preparations.[26] More importantly, selective 5-HT$_4$ receptor activation reversed an atropine-induced deficit in rat performance in the Morris Water Maze[18] and increased short-term olfactory memory (unpublished observations quoted in ref. 12). However, a reduction in short-term cognitive performance was not observed in rodents (delayed nonmatch to position test in the Skinner box) after dosing with brain-penetrant, selective 5-HT$_4$ receptor antagonists (D Rogers, unpublished) and there are no reports of similar enhancements of cognitive activities with either cisapride or metoclopramide in human subjects given these drugs. As such, the clinical significance of the observations in rodents must be treated with caution.

Other Possibilities

Emesis

5-HT$_4$ receptor antagonism may exert anti-emetic activity, although only against mild forms of emesis (induced by copper sulfate or 5-methoxytryptamine[4,24]) and not in all laboratories.[61] Since both the 5-HT$_3$ and 5-HT$_4$ receptors (in certain species) are located on vagal neurones, Gaster and Sanger[26] speculated that the higher affinity of the 5-HT$_4$ receptor for 5-HT, relative to the 5-HT$_3$ receptor, might mean that the former is more readily activated to evoke 'mild' forms of emesis, whereas higher amounts of 5-HT are required to activate the 5-HT$_3$ receptor and induce the severe emesis associated with many anti-cancer treatments. Nevertheless, an anti-emetic clinical indication for the 5-HT$_4$ receptor antagonists would not seem likely, given the availability of dopamine D$_2$ receptor antagonists for this type of emesis.[52]

Pain

The partial 5-HT$_4$ receptor agonists, BIMU1 or BIMU8, reduced thermal and mechanical nociception in rats and mice; this was prevented by SDZ 205-557 or by GR125487.[28] However, the more selective 5-HT$_4$ receptor antagonist RS 39604 had no effects against an index of visceral pain (increased arterial blood pressure) evoked in conscious rats by colorectal distension.[30] The latter was confirmed by Pichat et al[45] using normal rats, but these authors also reported on the ability of selective 5-HT$_4$ receptor antagonism to reduce the increased sensitivity to colorectal distension caused by mild wrap-restraint stress or acute irritation of the colorectum with acetic acid. If confirmed, this suggests that the antinociceptive activity of 5-HT$_4$ receptor antagonists is insufficiently robust to consider this class of agent as a general analgesic drug. However, if the antinociceptive properties of selective 5-HT$_4$ receptor antagonists are dependent on the medical condition studied, perhaps more work needs to be focused on its potential to reduce abdominal pain (e.g., Irritable Bowel Syndrome; see earlier).

Steroids

The therapeutic consequences of 5-HT$_4$ receptor-mediated stimulation of aldosterone and glucocorticoid secretion from the adrenal glands are not clear.[20] Whether or not 5-HT-induced cytokine release from the same tissues (rat adrenal zona glomerulosa; 5-HT released by stress)[49] is also mediated via the 5-HT$_4$ receptor requires further study.

Conclusions

In terms of 5-HT$_4$ receptor agonists, it is difficult to improve on the partial agonist cisapride, although new partial 5-HT$_4$ receptor agonists which also antagonize at the 5-HT$_3$ receptor may have some potential. Developing a full agonist at the receptor does not seem viable in terms of the high probability of creating unacceptable side-effects. As such, if significant advances are to be made in the drug treatment of gastrointestinal disorders, then a new approach, distinct from 5-HT$_4$ receptor agonism, must be considered. Early attempts are illustrated by the motilin receptor agonist erythromycin and by suggestions that inhibitors of nitric oxide synthase might remove an inhibitory influence on sphincter function.[59]

In terms of 5-HT$_4$ receptor antagonists, it is too soon to talk of combining this property with an ability to antagonize at a second receptor. Although this has been achieved with FK-1052 (for the treatment of IBS[46]), the combined effect inhibits intestinal propulsive activity[33] and may therefore induce constipation. Selective 5-HT$_4$ receptor antagonists are a new approach for the treatment of functional gastrointestinal disorders, acting to remove a cause of hypersensitivity without affecting the normal function of the gut. The first trial is being conducted by SmithKline Beecham, using the highly selective 5-HT$_4$ receptor antagonist SB-207266.[66]

References

1. Accarino AM, Azpiroz F, Malegelada J-R. Selective dysfunction of mechanosensitive intestinal afferents in irritable bowel syndrome. Gastroenterol 1995; 108:636-643.
2. Banner SE, Smith MI, Bywater D et al. Increased defecation caused by 5-HT$_4$ receptor activation in the mouse. Eur J Pharmacol 1996; 308:181-186.
3. Baxter GS, Craig DA, Clarke DE. 5-Hydroxytryptamine$_4$ receptors mediate relaxation of the rat oesophageal tunica muscularis mucosae. Naunyn-Schmiedeberg's Arch Pharmacol 1991; 343:439-446.
4. Bhandari P, Andrews PLR. Preliminary evidence for the involvement of the putative 5-HT$_4$ receptor in zacopride- and copper sulphate-induced vomiting in the ferret. Eur J Pharmacol 1991; 204:273-280.
5. Bijak M, Tokarski K, Maj J. Repeated treatment with antidepressant drugs induces subsensitivity to the excitatory effect of 5-HT$_4$ recep-

tor activation in the rat hippocampus. Naunyn-Schmiedeberg's Arch Pharmacol 1997; 355:14-19.

6. Bockaert J, Fozard JR, Dumuis A, Clarke DE. The 5-HT$_4$ receptor: a place in the sun. Trends Pharmacol Sci 1992; 13:141-145.

7. Brejer MR, Akkermans LMA, Schuurkes JAJ. Gastrointestinal prokinetic benzamides: The pharmacology underlying stimulation of motility. Pharmacol Rev 1995; 47:631-651.

8. Buchheit K-H, Gamse R, Giger R et al. The serotonin 5-HT$_4$ receptor. 1. Design of a new class of agonists and receptor map of the agonist recognition site. J Med Chem 1995; 38:2326-2330.

9. Burks TF, Long JP. Catecholamine-induced release of 5-hydroxytryptamine from perfused vasculature of isolated dog intestine. J Pharm Sci 1966; 55:1383-1386.

10. Candura SM, Messori E, Franceschetti GP et al. Neural 5-HT$_4$ receptors in the human isolated detrusor muscle: effects of indole, benzimidazolone and substituted benzamide agonists and antagonists. Br J Pharmacol 1996; 118:1965-1970.

11. Collins SM, McHugh K, Jacobson K et al. Previous inflammation alters the response of the rat colon to stress. Gastroenterol 1996; 111:1509-1515.

12. Compan V, Daszuta A, Salin P et al. Lesion study of the distribution of serotonin 5-HT$_4$ receptors in rat basal ganglia and hippocampus. Eur J Neurosci 1996; 8:2591-2598.

13. Consolo S, Arnaboldi S, Giorgi S et al. 5-HT$_4$ receptor stimulation facilitates acetylcholine release in rat frontal cortex. NeuroReport 1994; 5:1230-1232.

14. Costall B, Naylor RJ. The pharmacology of the 5-HT$_4$ receptor. Int Clin Psychopharmacol 1993; 8(suppl 2):11-18.

15. Costall B, Naylor RJ. 5-HT$_4$ receptor antagonists attenuate the disinhibitory effects of diazepam in the mouse light/dark test. Br J Pharmacol 1996; 119:352P.

16. De Deurwaerdere P, L'hirodel M, Bonhomme N et al. Serotonin stimulation of 5-HT$_4$ receptors indirectly enhances in vivo dopamine release in the rat striatum. J Neurochemistry 1997; 68:195-203.

17. Dumuis A, Bouhelal R, Sebben M et al. A non-classical 5-hydroxytryptamine receptor positively coupled with adenylate cyclase in the central nervous system. Mol Pharmacol 1988; 34:880-887.

18. Eglen RM, Bonhaus D, Clark RD et al. Effects of a selective and potent 5-HT$_4$ receptor agonist, RS-67333, and antagonist, RS-67532, in a rodent model of spatial learning and memory. Br J Pharmacol 1995; 116:235P.

19. Eglen RM, Swank SR, Walsh LKM, Whiting RL. Characterization of 5-HT$_3$ and 'atypical' 5-HT receptors mediating guinea-pig ileal contractions in vitro. Br J Pharmacol 1990; 101:513-520.

20. Eglen RM, Wong EHF, Dumuis A, Bockaert J. Central 5-HT$_4$ receptors. Trends Pharmacol Sci 1995; 16:391-398.

21. Elswood CJ, Bunce KT, Humphrey PPA. Identification of putative 5-HT$_4$ receptors in guinea-pig ascending colon. Eur J Pharmacol 1991; 1996:149-155.

22. Fargeas MJ, Fioromonto J, Bueno L. The role of monoaminergic systems in central IL-1beta-induced changes in intestinal motility. Neurogastroenterol Motility 1994; 6:189-195.

23. Ford APDW, Clarke DE. The 5-HT$_4$ receptor. Med Res Rev 1993; 13:633-662.

24. Fukui H, Yamamoto M, Sasaki S, Sato S. Possible involvement of peripheral 5-HT$_4$ receptors in copper sulfate-induced vomiting in dogs. Eur J Pharmacol 1994; 257:47-52.

25. Garter R. Management of anorexia-cachexia associated cancer and HIV-infection. Oncology 1991; 5 suppl:13-17.

26. Gaster LM, Sanger GJ. SB 204070: 5-HT$_4$ receptor antagonists and their potential therapeutic utility. Drugs of the Future 1994; 19:1109-1121.

27. Gerald C, Adham N, Kao H-T et al. The 5-HT$_4$ receptor: molecular cloning and pharmacological characterisation of two splice variants. EMBO J 1995; 14:2806-2815.

28. Ghelardini C, Galeotti N, Casamenti F et al. Central cholinergic antinociception induced by 5-HT$_4$ agonists: BIMU1 and BIMU8. Life Sci 1996; 58:2297-2309.

29. Hegde SS, Eglen RM. Peripheral 5-HT$_4$ receptors. FASEB J 1996; 10:1398-1407.

30. Hegde SS, Bonhaus DW, Johnson LG et al. RS 39604: a potent, selective and orally active 5-HT$_4$ receptor antagonist. Br J Pharmacol 1995; 115:1087-1095.

31. Hegde SS, Moy TM, Perry MR et al. Evidence for the involvement of 5-hydroxytryptamine 4 receptors in 5-hydroxytryptophan-induced diarrhea in mice. J Pharmacol Exp Ther 1994; 271:741-747.

32. Inman W, Kubota K. Tachycardia during cisapride treatment. Br Med J 1992; 305:1019-1020.

33. Kadowaki M, Wade PR, Gershon MD. Participation of 5-HT$_3$, 5-HT$_4$ and nicotinic receptors in the peristaltic reflex of guinea-pig distal colon. Am J Physiol 1996; 271 G849-G857.

34. Kaumann AJ. Do human atrial 5-HT$_4$ receptors mediate arrhythmias? Trends Pharmacol Sci 1994; 15:451-455.

35. Kaumann AJ, Sanders L, Brown AM et al. A 5-HT$_4$-like receptor in human right atrium. Naunyn-Schmiedeberg's Arch Pharmacol 1991; 344:150-159.

36. Kennett GA, Bright F, Trail B et al. Anxiolytic-like actions of the selective 5-HT$_4$ receptor antagonists SB-204070A and SB-207266A in rats. Neuropharmacol 1997; 36:707-712.

37. Lorrain J, Lebon F, Grosset A, O'Connor SE. Effects of zacopride and cisapride on piglet atrial 5-HT$_4$ receptors. Br J Pharmacol 1993; 108:251P.

38. Mayer EA, Gebhart GF. Basic and clinical aspects of visceral hyperalgesia. Gastroenterol 1994; 107:271-293.

39. Mayer EA, Munakata J, Mertz H et al. Visceral hyperalgesia and irritable bowel syndrome. In: Gebhart GF, ed. Visceral Pain, Progress in Pain Research and Management, Vol 5. Seattle: IASP Press, 1995:429-468.

40. Moore BA, Sharkey KA, Mantle M. Role of 5-HT in cholera toxin-induced mucin secretion in the rat small intestine. Am J Physiol 1996; 270:G1001-G1009.

41. Morley. Metoclopramide and cisapride to treat early satiety and anorexia due to nausea in the elderly. Drugs Aging 1996; 8:134-155.

42. Panocka I, Ciccocioppo R, Polidoric C et al. The 5-HT₄ receptor antagonist, GR113808, reduces ethanol intake in alchohol-prefering rats. Pharmacol Biochem Behav 1995; 52:255-259.

43. Patel S, Roberts J, Moorman J, Reavill C. Localization of serotonin-4 receptors in the striato-nigral pathway in rat brain. Neuroscience 1995; 69:1159-1167.

44. Pettersson G. The neuronal control of the serotonin content in mammalian enterochromaffin cells. Acta Physiol Scand 1979; 470(suppl):1-30.

45. Pichat P, Baudot X, Lechevalier P, Angel I. Inhibition by 5-HT₃ and 5-HT₄ antagonists of stress-induced colonic dysfunction and visceral pain in the rat. Gastroenterol 1996; 110:A735.

46. Prous J, Mealy N, Castaner J. FK-1052. Agent for irritable bowel syndrome. Drugs of the Future 1994; 19:1075-1077.

47. Read NW. Visceral afferent innervation and functional bowel disease: Evidence for dyssensation and altered reflex function. In: Mayer EA, Raybould HE, eds. Basic and Clinical Aspects of Chronic Abdominal Pain. Amsterdam: Elsevier Science, 1993:87-96.

48. Reavill C, Hatcher JP, Zetterstrom TSC et al. 5-HT₄ receptor antagonism does not affect dopamine-mediated behavioral effects in the rat. Br J Pharmacol 1996; 118:71P.

49. Ritchie PK, Knight HH, Ashby M, Judd AM. Serotonin increases interleukin-6 release and decreases tumor necrosis factor release from rat adrenal zona glomerulosa cells in vitro. Endocrine 1996; 5:291-297.

50. Roseto A, Cavaleri F, Debons-Guillemin MC et al. Human rotaviruses increase serotonin release from intestinal carcinoid cells cultured in vitro. Ann Virol 1981; 132E:383-388.

51. Rouzade ML, Fioramonti J, Bueno L. Role des recepteurs 5-HT₃ dans le controle par la cholecystokinine des relaxations transitoires du sphincter inferieur de l'oesophage chez le chien. Gastroenterol Clin Biol 1996; 20:575-580.

52. Sanger GJ. The pharmacology of anti-emetic agents. In: Andrews PLR, Sanger GJ, eds. Emesis in Anti-cancer therapy. Mechanisms and Treatment. Chapman & Hall Medical, New York: 1993:179-210.

53. Sanger GJ. Preclinical differences in 5-HT₃ receptor antagonist characteristics. In: Reynolds DJM, Andrews PLR, Davis CJ, eds. Serotonin and the Scientific Basis of Anti-emetic Therapy. Oxford Clinical Communications, Oxford: 1995:155-163.

54. Sanger GJ. 5-Hydroxytryptamine and functional bowel disorders. Neurogastroenterol Motil 1996; 8:319-331.
55. Sanger GJ, Gaster LM. 5-HT₄ receptor antagonists. Exp Opin Ther Patents 1994; 4:323-334.
56. Sempere AP, Duarte J, Cubezas C et al. Aggrevation of Parkinsonian tremor by cisapride. Clin Neuropharmacol 1995; 18:76-78.
57. Silverman DHS, Munakata JA, Ennes H et al. Regional cerebral activity in normal and pathological perception of visceral pain. Gastroenterol 1997; 112:64-72.
58. Silvestre JS, Fernandez AG, Palacious JM. Effects of 5-HT₄ receptor antagonists on rat behavior in the elevated plus-maze test. Eur J Pharmacol 1996; 309:219-222.
59. Tonini M. Recent advances in the pharmacology of gastrointestinal prokinetics. Pharmacol Res 1996; 33:217-226.
60. Tonini M, Candura SM. 5-HT₄ receptor agonists and bladder disorders. Trends in Pharmacol Sci 1996; 17:314.
61. Twissell DJ, Bountra C, Dale TJ et al. 5-HT₄ receptors are not involved in the emetic response to cisplatin, copper sulphate or R,S-zacopride in the ferret. Br J Pharmacol 1994; 113:22P.
62. Twycross R. The use of prokinetic drugs in palliative care. Eur J Palliative Care 1996; 2:143-145.
63. Uribe A, Alam M, Johansson O et al. Microflora modulates endocrine cells in the gastrointestinal mucosa of the rat. Gastroenterol 1994; 107:1259-1269.
64. Vilaro MT, Cortes R, Gerald C et al. Localization of 5-HT₄ receptor mRNA in rat brain by in situ hybridozation histochemistry. Mol Brain Res 1996; 43:356-360.
65. Wardle KA, Sanger GJ. The guinea-pig distal colon—a sensitive preparation for the investigation of 5-HT₄ receptor-mediated contractions. Br J Pharmacol 1993; 110:1593-1599.
66. Wardle KA, Bingham S, Ellis ES et al. Selective and functional 5-hydroxytryptamine₄ receptor antagonism by SB-207266. Br J Pharmacol 1996; 118:665-670.
67. Warner RRP, Feldman MG, Warner GM et al. Changes in blood serotonin concentration in mechanical obstruction of the small intestine. II. Findings in patients with intestinal obstruction. Surgery 1966; 59:758.
68. Zetterstrom TSC, Husum H, Smith S, Sharp T. Local application of 5-HT₄ antagonists inhibits potassium-stimulated GABA efflux from rat substantia nigra in vivo. Br J Pharmacol 1996,119:347P.

Index